T0257782

Mass Transfer: Numerical Analysis with Measurements

Mass Transfer: Numerical Analysis with Measurements

Edited by **Preston Runner**

LANRYE
INTERNATIONAL

New Jersey

Published by Clanrye International,
55 Van Reypen Street,
Jersey City, NJ 07306, USA
www.clanryeinternational.com

Mass Transfer: Numerical Analysis with Measurements
Edited by Preston Runner

International Standard Book Number: 978-1-63240-335-3 (Hardback)

Printed in the United States of America.

Contents

Preface

Every book is a source of knowledge and this one is no exception. The idea that led to the conceptualization of this book was the fact that the world is advancing rapidly; which makes it crucial to document the progress in every field. I am aware that a lot of data is already available, yet, there is a lot more to learn. Hence, I accepted the responsibility of editing this book and contributing my knowledge to the community.

In this book, experts provide recent developments in scientific findings and technologies, and introduce new theoretical models concerning mass transfer for sustainable energy and environment. The expertise of mass transfer processes has been extended and applied to different realm of science and engineering, including industrial applications in recent years. Since mass transfer is a primeval phenomenon, it plays a vital role in the scientific researches and fields of mechanical, energy, environmental, materials, bio and chemical engineering. The chapters have been grouped under a section advances in numerical analysis and measurement. The topics cover the developments in broad research areas, hence, the book will be informative not only to research engineers, but also to university professors and students.

While editing this book, I had multiple visions for it. Then I finally narrowed down to make every chapter a sole standing text explaining a particular topic, so that they can be used independently. However, the umbrella subject sinews them into a common theme. This makes the book a unique platform of knowledge.

I would like to give the major credit of this book to the experts from every corner of the world, who took the time to share their expertise with us. Also, I owe the completion of this book to the never-ending support of my family, who supported me throughout the project.

Editor

Advances in Numerical Analysis with Measurements

Study of Effect of Temperature Radient on Solid Dissolution Process Under Action of Transverse Rotating Magnetic Field

Rafał Rakoczy, Marian Kordas and Stanisław Masiuk

Additional information is available at the end of the chapter

1. Introduction

The design, scale-up and optimization of industrial processes conducted in agitated systems require, among other, precise knowledge of the hydrodynamics, mass and heat transfer parameters and reaction kinetics. Literature data available indicate that the mass-transfer process is generally the rate-limiting step in many industrial applications. Because of the tremendous importance of mass-transfer in engineering practice, a very large number of studies have determined mass-transfer coefficients both empirically and theoretically. From the practical point of view, the agitated systems are usually employed to dissolve granular or powdered solids into a liquid solvent [3].

Transfer of the solute into the main body of the fluid occurs in the three ways, dependent upon the conditions. For an infinite stagnant fluid, transfer will be by the molecular diffusion augmented by the gradients of temperature and pressure. The natural convection currents are set up owing to the difference in density between the pure solvent and the solution. This difference in inducted flow helps to carry solute away from the interface. The third mode of transport is depended on the external effects. In this way, the forced convection closely resembles natural convection expect that the liquid flow is involved by using the external force.

One of the key aspects in the dynamic behaviour of the mass-transfer processes is the role of hydrodynamics. On a macroscopic scale, the improvement of hydrodynamic conditions can be achieved by using various techniques of mixing, vibration, rotation, pulsation and oscillation in addition to other techniques like the use of fluidization, turbulence promotes or magnetic and electric fields etc. The transverse rotating magnetic field (TRMF) is a versatile

option for enhancing several physical and chemical processes. Studies over the recent decades were focused on application of magnetic field (MF) in different areas of engineering
processes [21, 22]. Static, rotating or alternating MFs might be used to augment the process
intensity instead of mechanically mixing. The practical applications of TRMF are presented
in the relevant literature [6, 16, 18, 23, 26, 27, 29].

Recently, TRMF are widely used to control different processes in the various engineering
operations [2, 9, 10]. This kind of magnetic field induces a time-averaged azimuthal force,
which drives the flow of the electrical conducting fluid in circumferential direction. According to available in technical literature, the mass-transfer during the solid dissolution to the
surrounding liquid under the action of TRMF has been deliberated [21, 22]. These papers
present literature survey for the applications of magnetic field (MF) and the magnetically assisted fluidization (MAF) in the mass transfer enhancement.

It should be noticed that the temperature gradient induces buoyancy-driven convective flow
in the fluid. This temperature gradient has a significant practical interest to the mass transfer
process. It is reported that the difference between the surface temperature of solid sample
and the liquid temperature has strong influence on the dissolution process [1].

The main objective of the present study is to investigate the solid dissolution process that is
induced under the action of TRMF and the gradient temperature between solid surface and
liquid. According to the information available in technical literature, the usage of TRMF and
gradient temperature is not theoretical and practical analyzed. The obtained experimental
data are generalized by using the empirical dimensionless correlations.

2. Theoretical background

2.1. Equation of magnetic induction

The flow under the action of TRMF may be determined by taking into consideration the following magnetohydrodynamic Ohm law

$$\frac{1}{\mu_m} rot \bar{B} = \bar{j}$$

(1)

The current density ($\bar{j}$) and the total electric field current ($\bar{E}$) may be expressed as follows

$$\bar{j} = \sigma_e \left[\bar{E} + \left(\bar{w} \times \bar{B} \right) \right]$$

(2)

$$\bar{E} = -\bar{w} \times \bar{B} + \frac{rot \bar{B}}{\sigma_e \mu_m}$$

(3)

The general Eq.(3) may be rewritten as:

$$rot\,\overline{E} = -rot\left(\overline{w} \times \overline{B}\right) + \frac{rot\,rot\,\overline{B}}{\sigma_e \mu_m} \tag{4}$$

Taking into consideration the following expressions

$$\Delta\overline{B} = -rot\,rot\,\overline{B} + grad\,div\,\overline{B} \tag{5}$$

$$rot\,\overline{E} = -\frac{\partial\overline{B}}{\partial\tau} \tag{6}$$

$$div\,\overline{B} = 0 \tag{7}$$

we obtain from the Eq.(3) the well-known advection-diffusion type relation [8, 17]

$$\frac{\partial\overline{B}}{\partial\tau} = rot\left(\overline{w} \times \overline{B}\right) + \frac{\Delta\overline{B}}{\sigma_e \mu_m} \tag{8}$$

The above equation (8) is also called the induction equation and it characterizes the temporal evolution of the magnetic field where

$$v_m = \frac{1}{\sigma_e \mu_m} \tag{9}$$

is effective diffusion coefficient (magnetic viscosity or magnetic diffusivity).

Taking into accunt the above relation, Eq. (8) may be revritten in the following form

$$\frac{\partial\overline{B}}{\partial\tau} = rot\left(\overline{w} \times \overline{B}\right) + v_m\Delta\overline{B} \tag{10}$$

The term, $rot\left(\overline{w} \times \overline{B}\right)$, in Eq.(8) dominates when the conductivity is large, and can be regarded as describing freezing of MF lines into the liquid. The term, $v_m\Delta\overline{B}$, in the B-field equation may be treated as a diffusion term. When the electrical conductivity, σ_e, is not too large, MF lines diffuse within the fluid.

Taking into account the below definitions of the dimensionless parameters

$$\overline{B}^* = \frac{\overline{B}}{B_0};\tau^* = \frac{\tau}{\tau_0};\overrightarrow{w}^* = \frac{\overline{w}}{w_0};l^* = \frac{l}{l_0};v_m^* = \frac{v_m}{v_{m_0}};rot^* = \frac{rot}{l_0^{-1}};\Delta^* = \frac{\Delta}{l_0^{-2}};\frac{\partial}{\partial\tau^*} = \frac{1}{\tau_0^{-1}}\frac{\partial}{\partial\tau} \tag{11}$$

we get the modified form of the relation (8)

$$\frac{B_0}{\tau_0}\left[\frac{\partial \vec{B}^*}{\partial \tau^*}\right] = \frac{w_0 B_0}{l_0}\left[rot^*\left(\vec{w}^* \times \vec{B}^*\right)\right] + \frac{\nu_{m_0} B_0}{l_0^2}\left[\Delta^* \vec{B}^*\right]$$
(12)

The above form of Eq. (8) may be used to examine the effect of liquid flow on the MF distribution. The non-dimensional forms of these equations may be scaled against the term $\left(\frac{\nu_{m_0} B_0}{l_0^2}\right)$. The dimensionless form of the equation (12) may be expressed by

$$\frac{l_0^2}{\nu_{m_0}\tau_0}\left[\frac{\partial \vec{B}^*}{\partial t^*}\right] = \frac{w_0 l_0}{\nu_{m_0}}\left[rot^*\left(\vec{w}^* \times \vec{B}^*\right)\right] + \left[\Delta^* \vec{B}^*\right]$$
(13)

This equation includes the following dimensionless groups

$$Fo_m = \frac{\nu_{m_0}\tau_0}{l_0^2}$$
(14)

and

$$Re_m = \frac{w_0 l_0}{\nu_{m_0}}$$
(15)

The magnetic Reynolds number (Re_m) is analogous to the traditional Reynolds number, describes the relative importance of advection and diffusion of the MF.

Taking into account the above definitions of the non-dimensional groups (Eqs (14) and (15)), we obtain the following general relationship of the magnetic induction equation

$$\frac{1}{Fo_m}\left[\frac{\partial \vec{B}^*}{\partial \tau^*}\right] = Re_m\left[rot^*\left(\vec{w}^* \times \vec{B}^*\right)\right] + \left[\Delta^* \vec{B}^*\right]$$
(16)

It should be noticed that the time of magnetic diffusion, τ_d, may defined as follows

$$\frac{1}{Fo_m} \sim 1 \Rightarrow \tau_0 \sim \frac{l_0^2}{\nu_{m_0}} \Rightarrow \tau_d = \sigma_e \mu_m l_0^2$$
(17)

Taking into account the following relation

$$rot\left(\overline{w} \times \overline{B}\right) \equiv \overline{B}\,grad\,\overline{w} - \overline{w}\,grad\,\overline{B} + \overline{w}\,div\,\overline{B} - \overline{B}\,div\,\overline{w} \tag{18}$$

we obtain the modified form of Eq. (8)

$$\overline{w}\,grad\,\overline{B} = \frac{\Delta \overline{B}}{\sigma_e \mu_m} \Rightarrow \overline{w}\,grad\,\overline{B} = \nu_m \Delta \overline{B} \tag{19}$$

The governing Eq.(19) may be rewritten in a symbolic shape which is useful in the case of the dimensionless analysis

$$\frac{w_0 B_0}{l_0}\left[\overrightarrow{w}^*\,grad^*\,\overline{B}^*\right] = \frac{\nu_{m_0} B_0}{l_0^2}\left[\nu_m^* \Delta^* \overline{B}^*\right] \tag{20}$$

The above Eq.(20) admits the following relation

$$\frac{w_0 B_0}{l_0} \sim \frac{\nu_{m_0} B_0}{l_0^2} \tag{21}$$

where $w_0 \equiv \omega_{TRMF}\,l_0$ and $\Delta^* = \dfrac{\Delta}{l_0^{-2}} \Rightarrow \Delta^* = \dfrac{\Delta}{\delta_0^{-2}}$.

An important consequence of the above expression (Eq.(21)) is that the skin depth (or the penetration depth), δ, may be given as follows

$$\frac{w_0 B_0}{l_0} \sim \frac{\nu_{m_0} B_0}{\delta_0^2} \Rightarrow \delta_0 \sim \sqrt{\frac{\nu_m l_0}{w_0}} \Rightarrow \delta = \sqrt{\frac{\nu_m}{\omega_{NPM}}} \tag{22}$$

This parameter may be used to describe the well-known skin effect. From the practical point of view, this phenomenon is characterized by the so-called shielding parameter

$$S = \frac{\omega_{NPM}\,l_0^2}{\nu_m} \tag{23}$$

This dimensionless parameter see (Eq (23)) is usually applied characterize the interaction between the MF and the electrical conductivity of liquid. The condition $S \ll 1$ means that MF is not changed by the conducting liquid. On the contrary, the condition $S \gg 1$ describes the typical skin effect which means that MF can penetrate into the highly electrically conductive liquid.

2.2. Influence of transverse rotating magnetic field on solid dissolution process

Under forced convective conditions, the mathematical description of the solid dissolution process may be described by means of the differential equation of mass balance for the component i

$$\frac{\partial \rho_i}{\partial \tau} + div\left(\rho_i \overline{w_i}\right) = \Phi_i \tag{24}$$

where Φ_i is the mass flux of component i (the volumetric mass source of component i).

The flux density of component i $\left(\overline{J_i}\right)$ may be given by

$$\overline{J_i} = \rho_i \overline{w_i} \tag{25}$$

The diffusion flux density is described by means of the following expression

$$\overline{J_{dyf}} = \rho_i \, dif \, \overline{w_i} \Rightarrow \overline{J_{dyf}} = \rho_i\left(\overline{w_i} - \overline{w}\right) \tag{26}$$

The relation between $\overline{J_i}$ and $\overline{J_{dyf}}$ is defined as

$$\overline{J_i} = \overline{J_{dyf}} + \rho_i \overline{w} \Rightarrow \rho_i \overline{w_i} = \rho_i\left(\overline{w_i} - \overline{w}\right) + \rho_i \overline{w} \tag{27}$$

Including the relation (27) in equation (24) gives the following relationship for the mass balance of component i

$$\frac{\partial \rho_i}{\partial \tau} + div\left[\overline{J_{dyf}}\right] + div\left(\rho_i \overline{w}\right) = \Phi_i \tag{28}$$

Introducing the relation

$$div\left(\rho_i \overline{w}\right) = \rho_i \, div\left(\overline{w}\right) + \overline{w} \, grad\left(\rho_i\right) \tag{29}$$

in Eq.(28) gives the mass balance of component i

$$\frac{\partial \rho_i}{\partial \tau} + \overline{w} \, grad\left(\rho_i\right) + \rho_i \, div\left(\overline{w}\right) + div\left[\overline{J_{dyf}}\right] = \Phi_i \tag{30}$$

The concentration of component i may be expressed as follows

$$c_i = \frac{\rho_i}{\rho} \Rightarrow \rho_i = \rho c_i \tag{31}$$

Taking into account the above equation, we find the modified form of Eq.(30)

$$\frac{\partial(\rho c_i)}{\partial \tau} + \overline{w}\, grad(\rho c_i) + (\rho c_i)\, div(\overline{w}) + div\left[\overline{J_{dyf}}\right] = \Phi_i \tag{32}$$

Eq.(32) may be rewritten by

$$\frac{\partial(\rho c_i)}{\partial \tau} + \overline{w}\, grad(\rho c_i) = \rho \frac{\partial c_i}{\partial \tau} + \rho \overline{w}\, grad(c_i) + c_i \frac{\partial \rho}{\partial \tau} + c_i \overline{w}\, grad(\rho) \tag{33}$$

or

$$\rho \frac{\partial c_i}{\partial \tau} + \rho \overline{w}\, grad(c_i) + c_i \left[\frac{\partial \rho}{\partial \tau} + \overline{w}\, grad(\rho) + \rho\, div(\overline{w})\right] + div\left[\overline{J_{dyf}}\right] = \Phi_i \tag{34}$$

The term in square brackets is so-called the continuity equation and this relation may be simplified in the following form

$$\frac{\partial \rho}{\partial \tau} + \overline{w}\, grad(\rho) + \rho\, div(\overline{w}) = 0 \Rightarrow \frac{d\rho}{d\tau} + div(\rho \overline{w}) = 0 \tag{35}$$

This leads to the final expression for the mass balance of component i:

$$\rho \frac{\partial c_i}{\partial \tau} + \rho \overline{w}\, grad(c_i) + div\left[\overline{J_{dyf}}\right] = \Phi_i \tag{36}$$

The total diffusion flux density $(\overline{J_{dyf}})$ is expressed as a sum of elementary fluxes considering the concentration($\overline{J}_i(c_i)$), temperature($\overline{J}_i(T)$), thermodynamic pressure gradient($\overline{J}_i(p)$), and the additional force interactions $\overline{J}_i(\overline{F})$ (e.g. forced convection as a result of fluid mixing) in the following form

$$\overline{J_{dyf}} = \overline{J}_i(c_i) + \overline{J}_i(T) + \overline{J}_i(p) + \overline{J}_i(\overline{F}) \tag{37}$$

A more useful form of this equation may be obtained by introducing the proper coefficients as follows

$$\overline{J_{dif}} = -\rho D_i\, grad(c_i) - \rho D_i k_t\, grad(\ln t) - \rho D_i k_p\, grad(\ln p) + \rho D_i k_{\overline{F}}\, \overline{F} \tag{38}$$

Under the action of TRMF the force $\overline{F}$ may be defined as the Lorenz magnetic force$\overline{F_{em}}$. This force is acting as the driving force for the liquid rotation and it may be described by

$$\overline{F_{em}} = \overline{J} \times \overline{B} \Rightarrow \overline{F_{em}} = \left[\sigma_e \overline{E} + \sigma_e\left(\overline{w} \times \overline{B}\right)\right] \times \overline{B} \tag{39}$$

The above relation may be simplified as follows (the electric field vector $\overline{E}$ is omitted)

$$\overline{F_{em}} = \left(\sigma_e\left(\overline{w} \times \overline{B}\right)\right) \times \overline{B} \tag{40}$$

The related Lorenz force to the unit of liquid mass may be rewritten in the form

$$\overline{F_{em}} = \frac{1}{\rho}\left(\sigma_e\left(\overline{w} \times \overline{B}\right)\right) \times \overline{B} \tag{41}$$

Introducing the relation Eq.(41) in Eq.(38) gives the following relationship

$$\overline{J_{dif}} = -\rho D_i\, grad(c_i) - \rho D_i k_t\, grad(\ln t) - \rho D_i k_p\, grad(\ln p) + D_i k_{\overline{F_{em}}}\left(\left(\sigma_e\left(\overline{w} \times \overline{B}\right)\right) \times \overline{B}\right) \tag{42}$$

Taking into account the above relation (Eq.(42)) we obtain the following general relationship for the mass balance of component i

$$\frac{\partial c_i}{\partial \tau} + \overline{w}\, grad(c_i) + div\left[-D_i\, grad(c_i) - D_i k_t\, grad(\ln t) - D_i k_p\, grad(\ln p)\right] +$$
$$+ div\left[\frac{D_i k_{\overline{F_{em}}}}{\rho}\left(\left(\sigma_e\left(\overline{w} \times \overline{B}\right)\right) \times \overline{B}\right)\right] = \frac{\Phi_i}{\rho} \tag{43}$$

The obtained Eq.(43) suggests that this dependence may be simplified in the following form

$$\frac{\partial c_i}{\partial \tau} + \overline{w}\, grad(c_i) + div\left[-D_i\, grad(c_i)\right] + div\left[\frac{D_m}{\rho}\left(\left(\sigma_e\left(\overline{w} \times \overline{B}\right)\right) \times \overline{B}\right)\right] = \frac{\Phi_i}{\rho} \tag{44}$$

It should be noticed that the coefficient of magnetic diffusion D_m may be expressed as follows

$$D_m = D_i k_{\overline{F_{em}}} \tag{45}$$

This coefficient may be defined by means of the following expression

$$D_m = c_t \tau_d \Rightarrow D_m = c_t \sigma_e \mu_m l_0^2 \Rightarrow D_m = \frac{c_t l_0^2}{v_m} \tag{46}$$

Taking into consideration the above relation (Eq.(46)) we obtain the relationship

$$\frac{\partial c_i}{\partial \tau} + \overline{w}\,grad(c_i) + div\left[-D_i\,grad(c_i)\right] + div\left[\frac{c_t l_0^2}{v_m \rho}\left(\left(\sigma_e\left(\overline{w}\times\overline{B}\right)\right)\times\overline{B}\right)\right] = \frac{\Phi_i}{\rho} \tag{47}$$

The above relation (Eq.(47)) may be treated as the differential mathematical model of the solid dissolution process under the action of TRMF. The right side of this equation represents the source mass of component i

$$\Phi_i = -\frac{\beta_i\,dF_m}{dV}\left(c_i - c_r\right) \Rightarrow \Phi_i = -\left(\beta_i\right)_V\left(c_i - c_r\right) \Rightarrow \Phi_i = -\left(\beta_i\right)_V \tilde{c}_i \Rightarrow \Phi_i = \left(\beta_i\right)_V\left(-\tilde{c}_i\right) \tag{48}$$

where $\left(-\tilde{c}_i\right)$ is the driving force for the solid dissolution process.

Introducing Eq.(48) in Eq.(47), gives the following relationship

$$\frac{\partial c_i}{\partial \tau} + \overline{w}\,grad(c_i) + div\left[-D_i\,grad(c_i)\right] + div\left[\frac{c_t l_0^2}{v_m \rho}\left(\left(\sigma_e\left(\overline{w}\times\overline{B}\right)\right)\times\overline{B}\right)\right] = -\frac{\left(\beta_i\right)_V \tilde{c}_i}{\rho} \tag{49}$$

Taking into account the below definition of the dimensionless parameters

$$c_i^* = \frac{c_i}{c_{i_0}}; \tilde{c}_i^* = \frac{\tilde{c}_i}{c_{i_0}}; \overline{w}^* = \frac{\overline{w}}{w_0}; D_i^* = \frac{D_i}{D_{i_0}}; \tau^* = \frac{\tau}{\tau_0}; \mu_m^* = \frac{\mu_m}{\mu_{m_0}};$$

$$v_m^* = \frac{v_m}{v_{m_0}}; \rho^* = \frac{\rho}{\rho_0}; \overline{B}^* = \frac{\overline{B}}{B_0}; \left[\left(\beta_i\right)_V\right]^* = \frac{\left(\beta_i\right)_V}{\left[\left(\beta_i\right)_V\right]_0} \tag{50}$$

$$div^* = \frac{div}{l_0^{-1}}; Div^* = \frac{Div}{l_0^{-1}}; grad^* = \frac{grad}{l_0^{-1}};$$

we obtain the governing Eq.(50) in a symbolic form

$$\frac{c_{i_0}}{\tau_0}\left[\frac{\partial c_i^*}{\partial \tau^*}\right] + \frac{c_{i_0}w_0}{l_0}\left[\overline{w}^*\,grad^*\left(c_i^*\right)\right] - \frac{D_{i_0}c_{i_0}}{l_0^2}\left[div^*\left[D_i^*\,grad^*\left(c_i^*\right)\right]\right] +$$

$$+ \frac{c_{i_0}\sigma_{e_0}w_0 B_0^2 l_0}{v_{m_0}\rho_0}\left[div^*\left[\frac{c_i^*}{v_m^*\rho^*}\left(\left(\sigma_e^*\left(\overline{w}^*\times\overline{B}^*\right)\right)\times\overline{B}^*\right)\right]\right] = -\frac{\left[\left(\beta_i\right)_V\right]_0 c_{i_0}}{\rho_0}\left[\frac{\left[\left(\beta_i\right)_V\right]^* \tilde{c}_i^*}{\rho^*}\right] \tag{51}$$

The non-dimensional form of this equation may be scaled against the convective term $\left(\dfrac{c_{i_0} w_0}{l_0}\right)$. The dimensionless form of Eq.(51) may be given as follows

$$\frac{l_0}{\tau_0 w_0}\left[\frac{\partial c_i^*}{\partial \tau^*}\right]+\left[\overrightarrow{w}^* grad^*\left(c_i^*\right)\right]-\frac{D_{i_0}}{l_0 w_0}\left[div^*\left[D_i^* grad^*\left(c_i^*\right)\right]\right]+$$
$$+\frac{\sigma_{e_0} B_0^2 l_0^2}{v_{m_0} \rho_0}\left[div^*\left[\frac{c_i^*}{v_m^* \rho^*}\left(\left(\sigma_e^*\left(\overrightarrow{w}^*\times\overrightarrow{B}^*\right)\right)\times\overrightarrow{B}^*\right)\right]\right]=-\frac{\left[(\beta_i)_v\right]_0 l_0}{\rho_0 w_0}\left[\frac{\left[(\beta_i)_v\right]^*\tilde{c}_i^*}{\rho^*}\right] \tag{52}$$

This relation includes the following dimensionless groups characterizing the dissolution process under the action of TRM

$$\frac{l_0}{\tau_0 w_0}\Rightarrow S^{-1} \tag{53}$$

$$\frac{D_{i_0}}{l_0 w_0}\Rightarrow\left(\frac{v}{w_0 D}\right)\left(\frac{D_{i_0}}{v}\right)\Rightarrow Re^{-1} Sc_i^{-1}\Rightarrow Pe_i^{-1} \tag{54}$$

$$\frac{\sigma_{e_0} B_0^2 l_0^2}{v_{m_0}\rho_0}\Rightarrow\left(\frac{\sigma_{e_0} B_0^2 l_0^2}{v\rho_0}\right)\left(\frac{v}{v_{m_0}}\right)\Rightarrow Q Pr_m\Rightarrow Ha^2 Pr_m \tag{55}$$

$$\frac{\left[(\beta_i)_v\right]_0 l_0}{\rho_0 w_0}\Rightarrow\left(\frac{\left[(\beta_i)_v\right]_0 d_p^2}{\rho_0 D_{i_0}}\right)\left(\frac{v}{w_0 D}\right)\left(\frac{D_{i_0}}{v}\right)\left(\frac{D^2}{d_s^2}\right)\Rightarrow Sh Sc^{-1} Re^{-1}\left(\frac{D^2}{d_s^2}\right) \tag{56}$$

Taking into account the proposed relations (53-56), we find the following dimensionless governing equation

$$S^{-1}\left[\frac{\partial c_i^*}{\partial \tau^*}\right]+\left[\overrightarrow{w}^* grad^*\left(c_i^*\right)\right]-Pe_i^{-1}\left[div^*\left[D_i^* grad^*\left(c_i^*\right)\right]\right]+$$
$$+Ha^2 Pr_m\left[div^*\left[\frac{c_i^*}{v_m^* \rho^*}\left(\left(\sigma_e^*\left(\overrightarrow{w}^*\times\overrightarrow{B}^*\right)\right)\times\overrightarrow{B}^*\right)\right]\right]=-Sh Sc^{-1} Re^{-1}\left(\frac{D^2}{d_s^2}\right)\left[\frac{\left[(\beta_i)_v\right]^*\tilde{c}_i^*}{\rho^*}\right] \tag{57}$$

From the dimensionless form of Eq.(57) it follows that

$$Sh Sc^{-1} Re^{-1}\left(\frac{D^2}{d_s^2}\right)\sim Ha^2 Pr_m\Rightarrow Sh Sc^{-1}\sim Ha^2 Re Pr_m\left(\frac{d_s^2}{D^2}\right)\Rightarrow Sh Sc^{-1}\sim Ta_m Pr_m\left(\frac{d_s^2}{D^2}\right) \tag{58}$$

Under convective conditions a relationship for the mass-transfer similar to the relationships obtained for heat-transfer may be expected of the form [11]

$$Sh = f(Re, Sc) \tag{59}$$

The two principle dimensionless groups of relevance to mass-transfer are Sherwood and Schmidt numbers. The Sherwood number can be viewed as describing the ratio of convective to diffusive transport, and finds its counterpart in heat transfer in the form of the Nusselt number [3].

The Schmidt number is a ratio of physical parameters pertinent to the system. This dimensionless group corresponds to the Prandtl number used in heat-transfer. Moreover, this number provides a measure of the relative effectiveness of momentum and mass transport by diffusion.

Added to these two groups is the Reynolds number, which represents the ratio of convective-to-viscous momentum transport. This number determines the existence of laminar or turbulent conditions of fluid flow. For small values of the Reynolds number, viscous forces are sufficiently large relative to inertia forces. But, with increasing the Reynolds number, viscous effects become progressively less important relative to inertia effects.

Evidently, for Eq.(59) to be of practical use, it must be rendered quantitative. This may be done by assuming that the functional relation is in the following form [4, 13]

$$Sh = a_1 Re^{b_1} Sc^{c_1} \tag{60}$$

The mass-transfer coefficients in the mixed systems can be correlated by the combination of Sherwood, Reynolds and Schmidt numbers. Using the proposed relation [60], it has been found possible to correlate a host of experimental data for a wide range of operations. The coefficients of relation [60] are determined from experiment. Under forced convection conditions the relation may be expressed as follows [7]

$$Sh \sim Re^{0.5} Sc^{0.33} \tag{61}$$

The exponent upon of the Schmidt number is to be 0.33 [5, 12, 14, 19, 25] as there is some theoretical and experimental evidence for this value [24], although reported values vary from 0.56 [28] to 1.13 [15].

Mass transfer process under the TRMF conditions is very complicated and may be described by the non-dimensional Eq.(57). Use of the dimensionless Sherwood number as a function of the various non-dimensional parameters yields a description of liquid-side mass transfer, which is more general and useful. Taking into account that the magnetic Prandtl number

$Pr_m = const$ (for water $Pr_m = const$) and the ratio of diameters solid sample and diameter of container $\dfrac{d_s}{D} = idem$, the obtained relationship (see Eq.(58)) may be expressed as follows

$$Sh\,Sc^{-1} \sim Ta_m Pr_m\left(\frac{d_s^2}{D^2}\right) \Rightarrow Sh = f\left(Ta_m, Sc\right) \tag{62}$$

Basing on the considerations given above, the correlations of mass transfer process under the TRMF action have the general form

$$Sh = a_2 Ta_m^{\ b_2} Sc^{c_2} \tag{63}$$

2.3. Influence of temperature gradient on the solid dissolution controlled process

As mentioned above, the temperature gradient has strong influence on the solid dissolution process. The heat transfer from the sample to the ambient fluid may be modeled by means of the well-known Nusselt type equation.

$$Nu = f\left(Re, Pr\right) \tag{64}$$

In the present report we consider the process dissolutions described by a similar but somewhat modified relationship between the dimensionless Sherwood number and the numbers which are defined the intensity of the magnetic effects in the tested experimental set-up with the TRMF generator. It should be assumed that the relationship for the heat transport under the TRMF conditions can be characterized in the following general form

$$Nu = f\left(Ta_m, Pr\right) \tag{65}$$

In order to establish the effect of all important parameters on this process in the wide range of variables data we proposed the following general relationship.

$$Nu = a_3 Ta_m^{\ b_3} Pr^{c_3} \tag{66}$$

The solid dissolution process under the action of TRMF and the gradient temperature between the solid surface and the liquid may be described by means of the following equations system

$$\begin{cases} Sh = a_2 Ta_m^{\ b_2} Sc^{c_2} \\ Nu = a_3 Ta_m^{\ b_3} Pr^{c_3} \end{cases} \tag{67}$$

From the above relation (Eq.(67)), the ratio of Sherwood and Nusselt numbers is given by

$$\frac{Sh}{Nu} = a_4 Ta_m^{b_4} \left(\frac{Sc}{Pr}\right)^{c_4} \tag{68}$$

where $a_2 \neq a_3 \wedge a_4 = \dfrac{a_2}{a_3}$; $b_2 \neq b_3 \wedge b_4 = b_2 - b_3$ and $c_2 = c_3 \wedge c_4 = 0.33$.

In the relevant literature the ratio of the Schmidt and Prandtl numbers is called as the dimensionless Lewis number (the ratio of thermal diffusivity to mass diffusivity)

$$Le = \left(\frac{Sc}{Pr}\right) \Rightarrow Le = \left(\frac{\nu}{D_i}\right)\left(\frac{a}{\nu}\right) \Rightarrow Le = \frac{a}{D_i} \tag{69}$$

According to Eq.(69) and the above assumptions, the ratio of Sherwood and Nusselt numbers is defined as follows

$$\frac{Sh}{Nu} = a_4 Ta_m^{b_4} Le^{0.33} \tag{70}$$

The enhancement effect of the solid dissolution process due to heat transfer process obtained from Eq.(70) is given by

$$\left(\frac{\beta_i d_s}{D_i}\right)\left(\frac{\lambda}{\alpha_s D}\right) = a_4 Ta_m^{b_4} Le^{0.33} \tag{71}$$

The ratio $\left(\dfrac{\beta_i}{\alpha_s}\right)$ may be defined by means of the following relationship

$$\left(\frac{\beta_i}{\alpha_s}\right) = a_4 Ta_m^{b_4} Le^{0.33} \left(\frac{D}{d_s}\right)\left(\frac{D_i}{\lambda}\right) \tag{72}$$

Introducing the thermal diffusivity $\left(a = \dfrac{\lambda}{c_p \rho} \Rightarrow \lambda = a c_p \rho\right)$ in Eq.(72), gives the relation

$$\left(\frac{\beta_i}{\alpha_s}\right) = a_4 Ta_m^{b_4} Le^{0.33} \left(\frac{D_i}{a}\right)\left(\frac{1}{c_p \rho}\right)\left(\frac{D}{d_s}\right) \Rightarrow \left(\frac{\beta_i}{\alpha_s}\right) = a_4 Ta_m^{b_4} Le^{0.33} Le^{-1} \left(\frac{1}{c_p \rho}\right)\left(\frac{D}{d_s}\right) \tag{73}$$

Finally, form relation (73) it follows that

$$\left(\frac{\beta_i}{\alpha_s}\right) = a_4 \mathrm{Ta}_m{}^{b_1} \mathrm{Le}^{-0.66}\left(\frac{1}{c_p\rho}\right)\left(\frac{D}{d_s}\right) \tag{74}$$

3. Experimental details

3.1. Experimental set-up

All experimental measurements of mass-transfer process using the TRMF were carried out in a laboratory set-up including electromagnetic field generator. A schematic of the experimental apparatus is presented in figure 1.

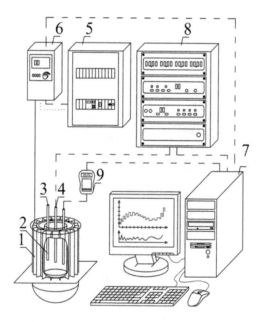

Figure 1. Sketch of experimental set-up: 1 - generator of rotating magnetic field, 2 - glass container, 3,4 - conductivity samples, 5 - electronic control box, 6 - a.c. transistorized inverter, 7 - personal computer, 8 - multifunctional electronic switch, 9 - Hall sample

This setup may be divided into: a generator of the rotating electromagnetic field (1), a glass container (2) with the conductivity samples (3-4), an electric control box (5) and an inverter (6) connected with multifunctional electronic switch (8) and a personal computer (7) loaded with special software. This software made possible the electromagnetic field rotation control, recording working parameters of the generator and various state parameters.

From preliminary tests of the experimental apparatus, the glass container is not influenced by the working parameters of the stator. The TRMF was generated by a modified 3-phase stator of an induction squirrel-cage motor, parameters of which are in accordance with the Polish Standard PN-72/E-06000. The stator is supplied with a 50 Hz three-phase alternating current. The transistorized inverter (4) was used to change the frequency of the rotating magnetic field in the range of $f_{TRMF} = 1 - 50 Hz$. The stator of the electric machine, as the RMF generator is made up of a number of stampings with slots to carry the three phase winding. The number of pair poles per phase winding, p, is equal to 2. The windings are geometrically spaced 120 degrees apart. The stator and the liquid may be treated as apparent virtual electrical circuit of the closed flux of a magnetic induction. The stator windings are connected through the a.c. transistorized inverter to the power source. The generator produces an azimuthal electromagnetic force in the bulk of the TRMF reactor with the magnetic field lines rotating in the horizontal plane.

For the experimental measurements, MF is generated by coils located axially around of the cylindrical container. As mentioned above, this field is rotated around the container with the constant angular frequency, ω_{TRMF}. The TRMF strength is determined by measuring a magnetic induction. The values of the magnetic induction at different points inside the glass container are detected by using a Hall sample connected to the personal computer. The typical example of the dependence between the spatial distributions of magnetic induction and the various values of the alternating current frequency for the cross-section of container is given [20]. The obtained results in this paper suggest that the averaged values of magnetic induction may be analytically described by the following relation

$$\left[B_{TRMF}\right]_{avg} = 14.05\left[1 - \exp\left(-0.05 f_{TRMF}\right)\right] \tag{75}$$

3.2. Rock-salt sample

Two conductive samples connected to a multifunction computer meter were used to measuring and recording of the concentration of the achieve solution of the salt. The mass of the rock salt sample decreasing during the process of dissolution is determined by an electronic balance that connected with rocking double-arm lever. On the lever arm the sample was hanging, the other arm connected to the balance. In the present investigation the change in mass of solid body in a short time period of dissolution is very small and the mean area of dissolved cylinder of the rock salt may be used. Than the mean mass-transfer Raw rock-salt (>98% NaCl and rest traces quantitative of chloride of K, Ca, Mg and insoluble mineral impurities) cylinders were not fit directly for the experiments because their structure was not homogeneous (certain porosity). Basic requirement concerning the experiments was creating possibly homogeneous transport conditions of mass on whole interfacial surface, which was the active surface of the solid body. These requirements were met thanks to proper preparing of the sample, mounting it in the mixer and matching proper time of dissolving. As an evident effect were fast showing big pinholes on the surface of the dissolved sample as results of local non-homogeneous of material. Departure from the shape of a simple geometrical body made it impossible to take measurements of its area with sufficient precision. So it was necessary to put those samples through the process of so-called hardening. The turned cylinders had been soaked in saturated brine solution for about 15 min and than dried in a room temperature. This process was repeated four times. To help mount the sample in the mixer, a thin copper thread was glued into the sample's axis. The processing was finished with additional smoothing of the surface with fine-grained abrasive paper. A sample prepared in this way had been keeping its shape during dissolving for about 30 min. The duration of a run was usually 30 sec. The rate of mass-transfer involved did not produce significant dimensional change in diameter of the cylinder. The time of a single dissolving cycle was chosen so that the measurement of mass loss could be made with sufficient accuracy and the decrease of dimensions would be relatively small (maximum about 0.5 mm).

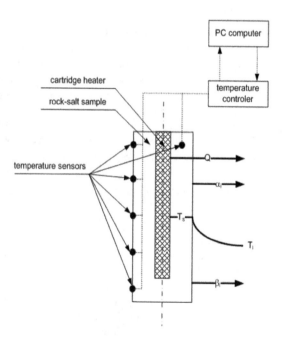

Figure 2. Sketch of rock-salt sample with the heating set-up

Before starting every experiment, a sample which height, diameter and mass had been known was mounted in a mixer under the free surface of the mixed liquid. The reciprocating plate agitator was started, the recording of concentration changes in time, the weight showing changes in sample's mass during the process of solution, and time measuring was started simultaneously. After finishing the cycle of dissolving, the agitator was stopped, and then the loss of mass had been read on electronic scale and concentration of NaCl (electrical conductivity) in the mixer as well. This connection is given by a calibration curve, showing the dependence of the relative mass concentration of NaCl on the electrical conductivity [21].

3.3. Rock-salt sample heated by means of the cartridge heater

In the case of this experimental investigations the gradient temperature between the surface and liquid was caused by using the cartridge heater (power ~1200W). This tubular device was inserted into drilled holes of rock-salt sample for heating. Moreover, the heating set-up was contained the temperature controller and sensors. The sensors for the temperature control was placed between the working surface of the sample and the heater. These sensors was also located on the surface of the solid sample. The sketch of rock-salt sample with the heating set-up is graphically presented in figure 2. The sample was kept at a constant temperature (65°C, 70°C or 80°C). The heat transfer from the sample to ambient fluid was realized for the various temperature (20°C, 40°C and 60°C). The system of temperature sensors was used to control the temperature of the water during the solid dissolution process.

3.4. Experimental calculation of mass transfer coefficient

The mass transfer coefficient under the action of TRMF may be calculated form the following equation

$$-\frac{dm_i}{d\tau} = \beta_i F_m \tilde{c}_i \Rightarrow \beta_i = \frac{1}{F_m \left(-\tilde{c}_i\right)} \frac{dm_i}{d\tau} \tag{76}$$

The above Eq.(76) cannot be integrated because the area of solid body, F_m, is changing in time of dissolving process. It should be noted that the change in mass of solid body in a short time period of dissolving is very small and the mean area of dissolved cylinder may be used. The relation between loss of mass, mean area of mass-transfer and the mean driving force of this process for the time of dissolving duration is approximately linear and then the mass-transfer coefficient may be calculated from the simple linear equation

$$\left[\beta_i\right]_{avg} = \frac{1}{\left[F_m\right]_{avg} \left[\tilde{c}_i\right]_{avg}} \frac{\left(-dif\, m_i\right)}{\left(dif\, \tau\right)} \tag{77}$$

The averaged surface F_m is defined as follows

$$\left[F_m\right]_{avg} = \pi \left[d_s\right]_{avg} h_s \tag{78}$$

The volumetric mass transfer coefficient $(\beta_i)_V$ in Eq.(48) is described by relation

$$(\beta_i)_V = \frac{\beta_e dF_m}{dV} \Rightarrow \left[(\beta_i)_V\right]_{avg} = \frac{\left[\beta_i\right]_r \left[F_m\right]_r}{V_l} \tag{79}$$

3.5. Experimental calculation of heat transfer coefficient

The heat transfer from the sample to liquid may by modelled by the following relationship

$$Q_s = Q_l \Rightarrow \alpha_s F_m \Delta T_1 = m_l c_{p_l} \Delta T_2 \tag{80}$$

This equation can be rewritten as

$$\alpha_s F_m \left(T_s - [T_l]_{t_2}\right) = m_l c_{p_l} \left([T_l]_{t_2} - [T_l]_{t_1}\right) \Rightarrow \alpha_s = \frac{m_l c_{p_l} \left([T_l]_{t_2} - [T_l]_{t_1}\right)}{F_m \left(T_s - [T_l]_{t_2}\right)} \tag{81}$$

and the averaged heat transfer coefficient is given as follows

$$[\alpha_s]_{avg} = \frac{m_l \left[c_{p_l}\right]_{avg} \left([T_l]_{t_2} - [T_l]_{t_1}\right)}{\left[F_m\right]_{avg} \left(T_s - [T_l]_{t_2}\right)} \tag{82}$$

The averaged coefficient of heat transfer $\left([\alpha_s]_{avg}\right)$ varies with the parameters of the TRMF mixing process and depends on the operating conditions and physical properties of the liquid.

4. Results and discussion

Under the TRMF conditions a relationship for the mass-transfer can be described in the general formSh$= f\left(Ta_m, Sc\right)$. The results of experiments suggest that the Sherwood number, the magnetic Taylor number and the Schmidt number may be defined as follows (see Eqs 53-56)

$$Sh = \frac{\left[(\beta_l)_V\right]_0 d_p^2}{\rho_0 D_{i_0}} \Rightarrow Sh = \frac{\left(\frac{[\beta_l]_{sr}[F_m]_{sr}}{V_l}\right) d_p^2}{\rho_l D_{i_l}} \tag{83}$$

$$Sc_i = \frac{v}{D_{i_0}} \Rightarrow Sc_i = \frac{v_l}{D_{i_l}} \tag{84}$$

$$Ta_m = Ha^2 Re_m \Rightarrow Ta_m = \left(\frac{\sigma_{e_0} B_0^2 l_0^2}{\nu \rho_0}\right)\left(\frac{w_0 D}{\nu}\right) \Rightarrow Ta_m = \left(\frac{\sigma_{e_l}\left([B_{TRMF}]_{avg}\right)^2 D^2}{\nu_l \rho_l}\right)\left(\frac{\omega_{TRMF} D^2}{\nu_l}\right)$$

$$\Rightarrow Ta_m = \frac{\omega_{TRMF}\left([B_{TRMF}]_{avg}\right)^2 D^4 \sigma_{e_l}}{\nu_l^2 \rho_l} \tag{85}$$

The TRMF Reynolds number $\left(Re_m = \dfrac{\omega_{TRMF} D^2}{\nu_l}\right)$ with $\omega_{TRMF} = 2\pi f_{TRMF}$ as angular frequency of TRMF equal to angular field frequency of the field generated by the a current of frequency. The product $\omega_{TRMF} D$ plays the role of a rotational velocity. The above dimensionless groups (Eqs 83-85) were calculated with the physical properties in the temperature range 20-60°C (the liquid temperature).

The effect of dissolution process under the action of TRF can be described by using the variable $ShSc^{-0.33}$ proportional to the term $a(Ta_m)^b$. The experimental results obtained in this work are graphically illustrated in $\log(ShSc^{-0.33})$ versus $\log(a(Ta_m)^b)$ in figure 3. Moreover, the influence of the temperature gradient between the surface temperature of solid and the liquid temperature on the mass transfer coefficient is presented in this figure.

In order to establish the effect of all important parameters on the dissolution process in the analyzed set-up, we propose the following relationship to work out the experimental database

$$\frac{Sh}{Sc^{0.33}} = a\left(Ta_m\right)^b \tag{86}$$

The presented results in figure 3 suggest that these points may be described by a unique monotonic function. The constants and exponents are computed by employing the Matlab software and the principle of least squares and the proposed relationships are collected in table 1.

Figure 4 shows the effect of the constant temperature of the surface of rock-salt sample and the variation of the liquid temperature on the Sherwood number.

temperature of surface of salt-rock sample	temperature of liquid		
	20°C	40°C	60°C
65°C	$\frac{Sh}{Sc^{0.33}} = 80.36\,(Ta_m)^{0.148}$	$\frac{Sh}{Sc^{0.33}} = 79.71\,(Ta_m)^{0.05}$	$\frac{Sh}{Sc^{0.33}} = 51.95\,(Ta_m)^{0.06}$
70°C	$\frac{Sh}{Sc^{0.33}} = 85.54\,(Ta_m)^{0.152}$	$\frac{Sh}{Sc^{0.33}} = 84.13\,(Ta_m)^{0.02}$	$\frac{Sh}{Sc^{0.33}} = 67.52\,(Ta_m)^{0.04}$
80°C	$\frac{Sh}{Sc^{0.33}} = 92.57\,(Ta_m)^{0.09}$	$\frac{Sh}{Sc^{0.33}} = 92.85\,(Ta_m)^{0.04}$	$\frac{Sh}{Sc^{0.33}} = 73.71\,(Ta_m)^{0.03}$

Table 1. The developed relationships for the obtained experimental data

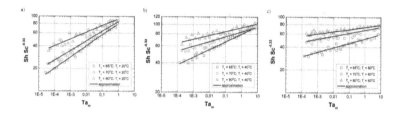

Figure 3. The graphical presentation of mass transfer data under the action of TRMF: a)T_s=var; T_l=20°C, b) T_s=var; T_l=40°Cand c) T_s=var; T_l=60°C

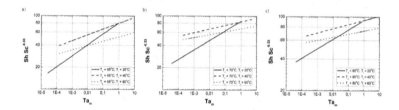

Figure 4. The comparison of obtained results: : a)T_s=65°C; T_l=var, b) T_s=70°C; T_l=varand c) T_s=80°C; T_l=var

Figures 3 and 4 present a graphical form of the collected relations in table 1, as the full curves, correlated the experimental data very well with the percentage relative error±10%. Figure 5 gives an overview results in the form of the proposed analytical relationships for the experimental investigations (see table 1)

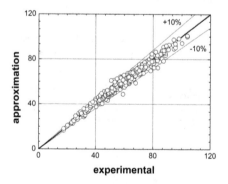

Figure 5. Dependence between experimental and predicted $\left(\dfrac{Sh}{Sc^{0.33}}\right)$ values

As can be clearly seen (see Figure 3) mass transfer rates expressed as $\left(\dfrac{Sh}{Sc^{0.33}}\right)$ increase with increasing the values of magnetic Taylor number. It is found that as the intensity of magnetic field increases, the velocity of liquid inside the cylindrical container increases. It may be concluded that the TRMF strongly influenced on the mass transfer process. It should be noticed that this process may be improved by means of the gradient temperature between the surface of rock-salt sample and the liquid. Figure 3 shows that Sherwood number increases with the increasing difference between the temperature of rock-salt surface and the liquid temperature. It is clear that the effect of TRMF on the dissolution process is also depended on the temperature gradient.

Comparison of the obtained results for the analyzed process is graphically presented in figure 4. This figure shows that for the given temperature of surface of rock-salt sample the mass transfer coefficients in tested set-up are strongly depended on the values of magnetic Taylor number. These plots also confirm that the gradient temperature has significant effect on the mass transfer process. Initially, the high mass transfer rates is achieved by the liquid temperature 40°C and 60°C. Further increase of the magnetic field intensity leads to even higher mass transfer rates for the liquid temperature is equal to 20°C. It should be noticed that the NaCl-cylinder was placed in the middle of container. When the TRMF rotated slowly the liquid was mixed near the wall of cylindrical container. When the TRMF rotated faster, the resulting liquid movement directly leads to an increase of the mass and heat transfer coefficients. This difference appears to be linked to the increase in the difference between the surface temperature of rock-salt sample and the liquid temperature associated with increasing the influence of TRMF. The high value of the exponents of magnetic Taylor number and the multiplicative coefficients seen in the relations given in table 1 agree with the existence of more intensive flow near the hot surface of the rock-salt sample promoted by the increase of the magnetic induction and the temperature of the cartridge heater.

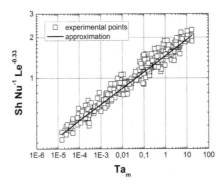

Figure 6. The graphical presentation of mass and heat transfer data at TRMF

The enhancement due to heat transfer process is modeled in terms given in Eq.(70). The graphical presentation of the calculated experimental points is presented in figure 6.

The constant a_4 and exponent b_4 in Eq.(70) are computed by using the principle of least square. Applying the software Matlab the analytical relationship may be obtained

$$\frac{Sh}{Nu} = 1.5\left(Ta_m\right)^{0.12} Le^{0.33} \tag{87}$$

where the ratio of the dimensionless Sherwood and Nusselt numbers is function of the adequate dimensionless groups. The fit of experimental data with Eq.(87) is given in figure 7. The averaged absolute relative error was estimated at 2.12%.

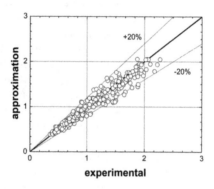

Figure 7. Comparison of model prediction (Eq.(87)) with experimental data

Figure 6 shows that the ratio of mass and heat transfer coefficients (via ratio of Sherwood and Nusselt numbers) increases with the magnetic Taylor number. This figure shows a strong increase in mass transfer process when the TRMF is applied. It was found that the intensification of this process is depended on the temperature gradient between the temperature of surface of salt-rock sample and the liquid temperature.

In order to evaluate the influence of the gradient temperature on the mass transfer under the action of TRMF, the comparison between the obtained database and the empirical correlation for the dissolution process under the TRMF is presented. For comparison these results with literature, it is recommended to correlate them under analogous form. The dissolution process under the action of TRMF is correlated by means of the equation [22]

$$Sh = 2 + 22.5\{Ta_m\}_x^{0.015} Sc^{0.33}\left(\frac{x}{D}\right)^{0.33} \tag{88}$$

Taking into account that the dimensionless location of a NaCl-cylindrical sample $\left(\dfrac{x}{D}\right)$ is equal to 0.125 (the sample was located in the middle of cylindrical container) and the local Taylor number $\left(\left[\text{Ta}_m\right]_x\right)$ is treated as the magnetic Taylor number (Ta_m), the Eq.(88) may be rewritten in the following form

$$\text{Sh} = 2 + 11.3 \left\{\text{Ta}_m\right\}_x^{0.015} \text{Sc}^{0.33} \Rightarrow \text{Sh}_I = 2 + 11.3 \left(\text{Ta}_m\right)^{0.015} \text{Sc}^{0.33} \tag{89}$$

As a matter of fact, Eq.(70) may be written by alternate equations as follows

$$\text{Sh}_{II} = 1.5 \left(\text{Ta}_m\right)^{0.12} \text{Le}^{0.33} \, \text{Nu} \tag{90}$$

The comparison in this case may be realized by considering the calculated averaged values of the dimensionless Schmidt $\left(\left[Sc\right]_{avg} = 477\right)$, Lewis $\left(\left[Le\right]_{avg} = 74\right)$ and Nusselt $\left(\left[Nu\right]_{avg} = 102\right)$ numbers. For established averaged values of these dimensionless groups the Eqs (89-90) reduce to

$$\text{Sh}_I = 2 + 86.5 \left(\text{Ta}_m\right)^{0.015} \tag{91}$$

$$\text{Sh}_{II} = 616.3 \left(\text{Ta}_m\right)^{0.12} \tag{92}$$

The graphical comparison between Eq.(91) and Eq.(92) are illustrated in the plot in figure 8. This figure demonstrates that the dimensionless Sherwood number for the analyzed case (Sh_{II}) increases with increasing the magnetic Taylor number. It was found that as the intensity of TRMF increases, the influence of hydrodynamic conditions on the transport processes inside the cylindrical container increases. The obtained relationship (Eqs (91-92)) indicate that the transfer rates increase with Taylor number for case I $\text{Sh}_I \sim (\text{Ta}_m)^{0.015}$ and case II $\text{Sh}_{II} \sim (Ta_m)^{0.12}$. The mass transfer data obtained for the additional transfer gradient is consequently higher than the data obtained for the mass transfer under the action of TRMF.

It can be observed that the enhancement of the mass transfer coefficients due to temperature gradient may be evaluated by applying the ratio $\left(\dfrac{\text{Sh}_{II}}{\text{Sh}_I}\right)$. In the present study $\left(\dfrac{\text{Sh}_{II}}{\text{Sh}_I}\right)$ becomes

$$\left(\frac{\text{Sh}_{II}}{\text{Sh}_I}\right) = \frac{616.3 \left(\text{Ta}_m\right)^{0.12}}{2 + 86.5 \left(\text{Ta}_m\right)^{0.015}} \Rightarrow \left(\frac{\text{Sh}_{II}}{\text{Sh}_I}\right) \approx 5 \left(\text{Ta}_m\right)^{0.105} \tag{93}$$

Figure 9 shows the obtained relation (see Eq.(93)) as the function of the magnetic Taylor number. It was found that as the intensity of TRMF has strong influence on the mass transfer rate. It is interesting to note that the enhancement of this process in the case of upper values of the magnetic Taylor number is increased for the supported process by using the cartrige heater.

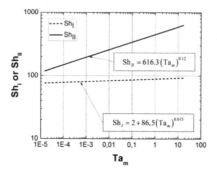

Figure 8. Comparison of obtained results (Sh_{II}) with literature data (Sh_I)

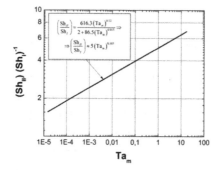

Figure 9. Graphical presentation of Eq.(93)

5. Conclusion

The present experimental study shows interestingfeatures cocncerning the effects of transverse rotating magnetic field (TRMF) on the mass transfer process. Inspecting the obtained measuremenst reveals the following conclusions:

1. The study of mass transfer process under the action of TRMF results is significant en-
    hancement of the solid dissolution rate per localization of a NaCl-cylindrical sample.
    The mass transfer rate increases with an increase of a magnetic field level. It was found
    that the TRMF strongly influcnecd the mass transfer process.

2. It should be noticed that the novel approach to the mixing process presented and based
    on the application of TRMF to produce better hydrodynamic conditions in the case of
    the mass-transfer process. From practical point of view, the dissolution process of solid
    body is involved by using the turbulently agitated systems. In previous publications are
    not available data describing the mass-transfer operations of the dissolution process un-
    der the TRMF conditions and the temperature gradient. Moreover, the influence of the
    additional indirect heating on the mass-transfer was determined.

3. With the respect to the other very useful mass transfer equations given in the pertinent
    literature, the theoretical description of problem and the equations predicted in the
    present article is much more attractive because it generalizes the experimental data tak-
    ing into consideration the various parameters, which defined the hydrodynamic state
    and the intensity of magnetic effects in the tested system (see chapter 2.2).

4. On the basis of the experimental investigations, the results were successfully correlated
    by using the general relationship [74]. The influence of the TRMF and the temperature
    gradient on this process may be also described using the non-dimensional parameters
    formulated on the base of fluid mechanics equations. These dimensionless numbers al-
    low quantitative representation and characterization of the influence of hydrodynamic
    state under the TRMF conditions on the mass-transfer process. The dimensionless
    groups are used to establish the effect of TRMF on this operation in the form of the nov-
    el type dimensionless correlation [87].

5. In order to evaluate the influence of the gradient temperature on the mass transfer un-
    der the action of TRMF, the comparison between the obtained database and the empiri-
    cal correlation for the dissolution process under the TRMF is presented. This coparsion
    is presented as the specific relation [93]. It can be observed that the enhancement of the
    mass transfer coefficients due to temperature has strong influence on the mass transfer
    rate (see Fig.8).

6. Nomenclature

$\bar{B}$	magnetic induction	$kg \cdot A^{-1} \cdot s^{-2}$
c_i	concentration	$kg_i \cdot kg^{-1}$
c_p	specific heat capacity of liquid	$J \cdot kg^{-1} \cdot deg^{-1}$
d_s	sample diameter	m

D	diameter of container	m
D_i	diffusion coefficient	$m^2 \cdot s^{-1}$
D_m	magnetic diffusion	$m^2 \cdot s^{-1}$
$\dot{E}$	electric field	$V \cdot m^{-1}$
f_{TRMF}	frequency of electrical current (equal to frequency of TRMF)	s^{-1}
F_m	cylindrical surface of dissoluble sample	m^2
$\overline{F_{em}}$	Lorenz magnetic force	N
h_s	length of sample	m
l	characteristic dimension	m
$\overline{J}$	electrical current density vector	$A \cdot m^{-2}$
$\overline{J}_i$	flux density of component i	$kg_i \cdot m^{-3} \cdot s^{-1}$
$\overline{J_{dyf}}$	diffusion flux density	$kg_i \cdot m^{-3} \cdot s^{-1}$
k_p	relative coefficient of barodiffusion	$kg_i \cdot m^2 \cdot kg^{-1} \cdot N^{-1}$ $\left(kg_i \cdot m \cdot s^2 \cdot kg^{-2}\right)$
k_p	relative coefficient of thermodiffusion	$kg_i \cdot kg^{-1} \cdot deg^{-1}$
$k_{F_{em}^-}$	relative coefficient of diffusion resulting from additional forced interactions (e.g. magnetic field)	$kg_i \cdot m^{-1} \cdot N^{-1}$ $\left(kg_i \cdot s^2 \cdot m^{-2} \cdot kg^{-1}\right)$
m_i	mass of dissoluble NaCl sample	kg_{NaCl}
p	hydrodynamic pressure	$N \cdot m^{-2}$
Q_l	heat flow from liquid	W
Q_s	heat flow from sample	W
S	shielding parameter	-
T	temperature	deg
$[T_l]_{t_1}$	temperature of liquid at moment t_1	deg
$[T_l]_{t_2}$	temperature of liquid at moment t_2 (after time of dissolution process)	deg
T_s	temperature of sample	deg
V	volume of liquid	m^3
$\bar{w}$	velocity	$m \cdot s^{-1}$
$\bar{w}_i$	velocity of component i	$m \cdot s^{-1}$
x	distance (for localization of sample)	m

Table 2.

a_s	heat transfer coefficient	$W \cdot m^{-2} \cdot deg^{-1}$
β_i	mass transfer coefficient	$kg_i \cdot m^{-2} \cdot s^{-1}$
δ	skin depth	-
η	dynamic viscosity	$kg \cdot m^{-1} \cdot s^{-1}$
μ_m	magnetic permeability	$kg \cdot m \cdot A^{-2} \cdot s^{-2}$
λ	thermal conductivity of liquid	$W \cdot m^{-1} \cdot deg^{-1}$
v	kinematic viscosity	$m^2 \cdot s^{-1}$
v_m	magnetic viscosity	$m^2 \cdot s^{-1}$
ρ	density	$kg \cdot m^{-3}$
ρ_i	concentration of component i	$kg_i \cdot m^{-3}$
σ_e	electrical conductivity	$A^2 \cdot s^3 \cdot kg^{-1} \cdot m^{-3}$
τ	time dissolution or time	s
Φ_i	mass flux of component i	$kg_i \cdot m^{-3} \cdot s^{-1}$
ω_{TRMF}	angular velocity of transverse rotating magnetic field	$rad \cdot s^{-1}$

Greek letters

avg	averaged value
l	liquid
s	sample
0	reference value

Subscripts

AC	alternating current
MF	magnetic field
TRMF	transverse rotating magnetic field

Abbreviation

$Ha = B_0 l_0 \sqrt{\dfrac{\sigma_{e_0}}{v \rho_0}}$	Hartman number

$Le = \dfrac{a}{D_i}$	Lewis number
$Nu = \dfrac{[a_s]_{avg} D}{\lambda}$	Nusselt number
$Pe_i = \dfrac{l_0 w_0}{D_{i_0}}$	mass Peclet number
$Pr_m = \dfrac{v}{v_{m_0}}$	magnetic Prandtl number
$Q = \dfrac{\sigma_{e_0} B_0^2 l_0^2}{v \rho_0}$	Chandrasekhar number
$Re = \dfrac{w_0 D}{v}$	Reynols number
$S = \dfrac{\tau_0 w_0}{l_0}$	Strouhal number
$Sc_i = \dfrac{v}{D_{i_0}}$	Schmidt number
$Sh = \dfrac{[(\beta_i)_V]_0 d_p^2}{\rho_0 D_{i_0}}$	Sherwood number

Dimensionless numbers

Acknowledgements

This work was supported by the Polish Ministry of Science and Higher Education from sources for science in the years 2012-2013 under Inventus Plus project

Author details

Rafał Rakoczy*, Marian Kordas and Stanisław Masiuk

*Address all correspondence to: rrakoczy@zut.edu.pl

Institute of Chemical Engineering and Environmental Protection Process, West Pomeranian University of Technology, Poland

References

[1] Aksielrud, G. A., & Mołczanow, A. D. (1981). *Dissolution process of solid bodies*, WNT Poland (in polish).

[2] Al-Qodah, Z., Al-Bisoul, M., & Al-Hassan, M. (2001). Hydro-thermal behavior of magnetically stabilized fluidized beds. *Powder Technology*, 115, 58-67.

[3] Basmadjian, D. (2004). *Mass transferPrinciples and applications* CRC Press LLC, USA.

[4] Bird, R. B., Stewart, W. E., & Lightfoot, E. N. (1966). *Transport phenomena*, Wiley, USA.

[5] Condoret, J. S., Riba, J. P., & Angelino, H. (1989). Mass transfer in a particle bed with oscillating flow. *Chemical Engineering Science*, 44(10), 2107-2111.

[6] Fraňa, K., Stiller, J., & Grundmann, R. (2006). Transitional and turbulent flows driven by a rotating magnetic field. *Magnetohydrodynamics*, 42, 187-197.

[7] Garner, F. H., & Suckling, R. D. (1958). Mass transfer form a soluble solid sphere. *AIChE Journal*, 4(1), 114-124.

[8] Guru, B. S., & Hiziroğlu, H. R. (2004). *Electromagnetic field theory fundamentals*, Cambridge University Press.

[9] Hristov, J. (2003). Magnetic field assisted fluidization- A unified approach. Part 3: Heat transfer in gas-solid fluidized beds- a critical re-evaluation of the results. [a], *Reviews in Chemical Engineering*, 19(3), 229-355.

[10] Hristov, J. (2003). Magnetic field assisted fluidization- A unified approach. Part 7: Mass Transfer: Chemical reactors, basic studies and practical implementations thereof. [b], *Reviews in Chemical Engineering*, 25(1-3), 1-254.

[11] Incropera, F. P., & De Witt, D. P. (1996). Fundamentals of heat and mass transfer. John Wiley & Sons Inc., USA.

[12] Jameson, G. J. (1964). Mass (or heat) transfer form an oscillating cylinder. *Chemical Engineering Science*, 19, 793-800.

[13] Kays, W. M., & Crawford, M. E. (1980). Convective heat and mass transfer. McGraw-Hill, USA.

[14] Lemcoff, N. O., & Jameson, G. J. (1975). Solid-liquid mass transfer in a resonant bubble contractor. *Chemical Engineering Science*, 30, 363-367.

[15] Lemlich, R., & Levy, M. R. (1961). The effect of vibration on natural convective mass transfer. *AIChE Journal*, 7, 240-241.

[16] Melle, S., Calderon, O. G., Fuller, G. G., & Rubio, M. A. (2002). Polarizable particle aggregation under rotating magnetic fields using scattering dichroism. *Journal of Colloid and Interface Science*, 247, 200-209.

[17] Mößner, R., & Gerbeth, G. (1999). Buoyant melt flows under the influence of steady and rotating magnetic fields. *Journal of Crystal Growth*, 197, 341-345.

[18] Nikrityuk, P. A., Eckert, K., & Grundmann, R. (2006). A numerical study of unidirectional solidification of a binary metal alloy under influence of a rotating magnetic field. *International Journal of Heat and Mass Transfer*, 49, 1501-1515.

[19] Noordsij, P., & Rotte, J. W. (1967). Mass transfer coefficients to a rotating and to a vibrating sphere. *Chemical Engineering Science*, 22, 1475-1481.

[20] Rakoczy, R., & Masiuk, S. (2009). Experimental study of bubble size distribution in a liquid column exposed to a rotating magnetic field. *Chemical Engineering and Processing: Process Intensification*, 48, 1229-1240.

[21] Rakoczy, R., & Masiuk, S. (2010). Influence of transverse rotating magnetic field on enhancement of solid dissolution process. *AIChE Journal*, 56, 1416-1433.

[22] Rakoczy, R. (2010). Enhancement of solid dissolution process under the influence of rotating magnetic field. *Chemical Engineering and Processing: Process intensification*, 49, 42-50.

[23] Spitzer, K. H. (1999). Application of rotating magnetic field in Czochralski crystal growth. *Crystal Growth and Characterization of Materials*, 38, 39-58.

[24] Sugano, Y., & Rutkowsky, D. A. (1968). Effect of transverse vibration upon the rate of mass transfer for horizontal cylinder. *Chemical Engineering Science*, 23, 707-716.

[25] Tojo, K., Miyanami, K., & Mitsui, H. (1981). Vibratory agitation in solid-liquid mixing. *Chemical Engineering Science*, 36, 279-284.

[26] Volz, M. P., & Mazuruk, K. (1999). Thermoconvective instability in a rotating magnetic field. *International Journal of Heat and Mass Transfer*, 42, 1037-1045.

[27] Walker, J. S., Volz, M. P., & Mazuruk, K. (2004). Rayleigh-Bénard instability in a vertical cylinder with a rotating magnetic field. *International Journal of Heat and Mass Transfer*, 47, 1877-1887.

[28] Wong, P. F. Y., Ko, N. W. M., & Yip, P. C. (1978). Mass transfer from large diameter vibrating cylinder. *Trans. Item*, 56, 214-216.

[29] Yang, M., Ma, N., Bliss, D. F., & Bryant, G. G. (2007). Melt motion during liquid-encapsulated Czochralski crystal growth in steady and rotating magnetic field. *International Journal of Heat and Mass Transfer*, 28, 768-776.

Chapter 2

Convective Mass Transfer in a Champagne Glass

Fabien Beaumont, Gérard Liger-Belair and
Guillaume Polidori

Additional information is available at the end of the chapter

1. Introduction

Legend has it that the Benedictine monk Dom Pierre Pérignon discovered the Champagne method for making sparkling wines more than 300 years ago. As it happens, a paper presented to the Royal Society in London described the Champagne production method in 1662, six years before Pérignon ever set foot in a monastery. In fact, Pérignon was first tasked with keeping bubbles out of wine, as the effervescence was seen as vulgar at the time. But then tastes changed and fizz became fashionable, so Pérignon's mandate was reversed; he went on to develop many advances in Champagne production, including ways to increase carbonation. In any case, the process was not regularly used in the Champagne region of France to produce sparkling wine until the 19th century. Since that time, Champagne has remained the wine of celebration, undoubtedly because of its bubbling behavior.

But what is the exact role of the bubbles? Is it just aesthetics? Do they contribute to only one aspect, or to many aspects, of the subjective final taste? We have been rigorously analyzing Champagne for more than a decade, using the physics of fluids in the service of wine in general and Champagne-tasting science in particular.

2. The Champagne method

Fine sparkling wines and Champagne result from a two-step fermentation process. After completion of the first alcoholic fermentation, some flat Champagne wine (called base wine) is bottled with a mixture of yeast and sugar. Consequently, a second fermentation starts inside the bottle as the yeast consumes the sugar, producing alcohol and a large amount of carbon dioxide (CO_2). This is why Champagne has a high concentration of CO_2 dissolved in it about 10 grams per liter of fluid and the finished Champagne wine can be under as much as five or six atmospheres of pressure. As the bottle is opened, the gas gushes out in the form of tiny CO_2 bubbles. In order for the liquid to regain equilibrium once the cork is removed, it must release about five liters of CO_2 from a 0.75 liter bottle, or about six times its own volume. About 80 percent of this CO_2 is simply outgassed by direct diffusion, but the remaining 20 percent still equates to about 20 million bubbles per glass (a typical flute holds about 0.1 liter). For Champagne connoisseurs, smaller bubble size is also a measure of quality.For consumers and winemakers as well, the role usually ascribed to bubbles in Champagne tasting is to awaken the sight sense. Indeed, the image of Champagne is intrinsically linked to the bubbles that look like "chains of pearls" in the glass and create a cushion of foam on the surface. But beyond this visual aspect, the informed consumer recognizes effervescence as one of the main ways that flavor is imparted, because bursting CO_2 bubbles propel the aroma of sparkling wine into the drinker's nose and mouth (Figure 1).

Figure 1. A glass of Champagne is a feast for all the senses; indeed it is sight and sound that make sparkling wines particularly special. Elegant bubble trains rise from nucleation sites suspended in the fluid *(right)*. Bubbles reaching the top of the glass burst and produce a fog of droplets *(above)*. The questions being explored by enologists include how the carbonation and effervescence induce fluid flow in, and affect the flavor of, the beverage. (All photographs are courtesy of the authors.)

One cannot understand the bubbling and aromatic exhalation events in champagne tasting, however, without studying the flow-mixing mechanisms inside the glass. Indeed, a key assumption is that a link of causality may exist between flow structures created in the wine due to bubble motion and the process of flavor exhalation [1]. But the consequences of the bubble behavior on the dynamics of the Champagne inside the glass and the CO_2 propelling process are still unknown. Quantifying the exhalation of flavors and aromas seems a considerable challenge, something that is difficult to control experimentally, but this constitutes the aim of our current work.

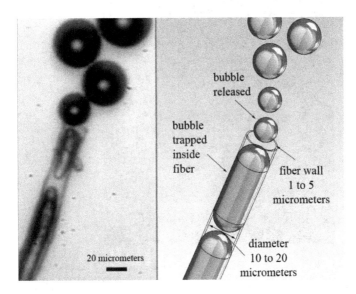

Figure 2. Bubbles in sparkling wines do not spring into existence unaided, but require a starting point. These nucleation sites take the form of microscopic cellulose fibers, from the air or a towel used to dry the glass, which trap air pockets as the glass is filled. Carbon dioxide from the wine diffuses into the gas pockets, producing bubbles like clockwork *(left).*

3. The birth of bubbles

The first step is to elucidate how bubbles themselves come into being. Generally speaking, two ways exist, and sometimes coexist, to generate bubble chains in Champagne glasses [2-6]. Natural effervescence depends on a random condition: the presence of tiny cellulose fibers deposited from the air or left over after wiping the glass with a towel, which cling to the glass due to electrostatic forces (Figure 2). These fibers are made of closely packed microfibrils, themselves consisting of long polymer chains composed mainly of glucose. Each fiber, about 100 micrometers long, develops an internal gas pocket as the glass is filled. Capillary action tries to pull the fluid inside the micro-channel of the fiber, but if the fiber is completely submerged before it can be filled, it will hold onto its trapped air. Such gas trapping is aided when the fibers are long and thin, and when the liquid has a low surface tension and high viscosity. Champagne has a surface tension about 30 percent less than that of water, and a viscosity about 50 percent higher.

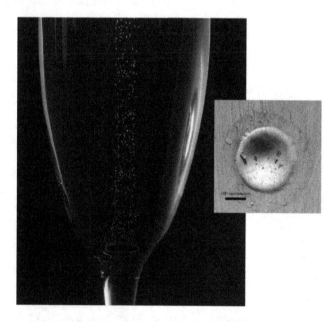

Figure 3. In order to study effervescence in Champagne and other sparkling wines, random bubble production must be replaced with controlled creation of bubble streams. The glass bottom is etched with a ring that provides nucleation sites for regular bubble trains *(left)*. The ring consists of many small impact points *(right)* from a laser, one of which is shown above. Glasses etched with a single nucleation point were used in studies to see how a single stream of bubbles would induce motion in the surrounding fluid, and what shape that fluid motion would take.

These microfiber gas pockets act as nucleation sites for the formation of bubbles. To aggregate, CO_2 has to push through liquid molecules held together by van der Waals forces, which it would not have enough energy to do on its own. The gas pockets lower the energy barrier to bubble formation (as long as they are above a critical size of 2 micrometers in radius, because below that size the gas pressure inside the bubble is too high to permit CO_2 to diffuse inside). It should be noted that irregularities in the glass surface itself cannot act as nucleation sites such imperfections are far too small, unless larger micro scratches are purposely made. Once a bubble grows to a size of 10 to 50 micrometers, it is buoyant enough to detach from the fiber, and another one forms like clockwork; an average of 30 bubbles per second are released from each fiber. The bubbles expand from further diffusion of CO_2 into them as they rise, which increases their buoyancy and accelerates their speed of ascent [2, 5-6]. They usually max out at less than a millimeter in diameter over the course of their one-to five-second travel time up the length of a flute. Because natural nucleation is very random and not easily controllable, another way to generate bubbles is to use a mechanical process that is perfectly reproducible from one filling to the next. Glassmakers use a laser to engrave artificial nucleation sites at the bottom of the glass; such modified glasses are commonly used by Champagne houses during tastings (Figure 3). To make the effervescence pattern pleasing to the eye, artisans use no fewer than 20 impacts to create a ring shape, which produces a regular column of rising bubbles [3-5, 7].

4. Fizz and flow

The displacement of an object in a quiescent fluid induces the motion of fluid layers in its vicinity. Champagne bubbles are no exception to this rule, acting like objects in motion, no matter whether the method used to produce them was random or artificial. Viscous effects make the lower part of a bubble a low-pressure area, which attracts fluid molecules around it and drags some fluid to the top surface, although the bubbles move about 10 times faster than the fluid (Figure 4). Consequently, bubbles and their neighboring liquid move as concurrent upward flows along the center line of the glass. Because the bubble generation from nucleation sites is continuous, and because a glass of Champagne is a confined vessel, this constant upward ascent of the fluid ineluctably induces a rotational flow as well [2-5, 7-8]. To get a precise idea of the role bubbles play in the fluid motion, we observed a Champagne flute with single nucleation site at the bottom (Figure 4). A bubble's geometric evolution is well studied in carbonated beverages. For example, we know that the bubble growth rate during vertical ascent reliably leads to an average diameter of about 500 micrometers for a 10 centimeter migration length in a flute. In fact, for such a liquid supersaturated with dissolved CO_2 gas molecules, empirical relationships reveal the bubble diameter to be proportional to the cube root of the vertical displacement. Another property of bubbles is that they can act as either rigid or flexible spheres as they rise, depending on the content of the fluid they are in, and rigid spheres experience more drag than flexible ones. Champagne bubbles do not act as rigid spheres, whereas bubbles in other fizzy fluids, such as beer, do. Beer contains a lot of proteins, which coat the outside of the bubbles as they ascend, preventing their deformation. Beer is also less carbonated than Champagne, so bubbles in it do not grow as quickly, making it easier for proteins to completely encircle them. But Champagne is a relatively low-protein fluid, so there are fewer surfactants to stick to the bubbles and slow them down as they ascend. In addition, Champagne's high carbonation makes bubbles grow rapidly on their upwards trip, creating ever more untainted surface area, in effect cleaning themselves of surfactants faster than new molecules can fill in the space. However, some surfactants are necessary to keep bubbles in linear streams with none; fluid flows would jostle the bubbles out of their orderly lines.

We carried out filling experiments at room temperature to avoid condensation on the glass surface, and allowed the filled glass to settle for a minute or so before taking measurements [2-5, 7-9]. Our visualization is based on a laser tomography technique, where a laser sheet 2 millimeters wide crosses the center line of the flute, imaging just this two- dimensional section of the glass using long-exposure photography. To avoid optical distortions by the curved surface of the glass, this latter is partially immersed in a parallelepiped tank full of water (Figure 5) with a refractive index close to that of champagne (RI champagne 1.342 while RI pure water 1.332).

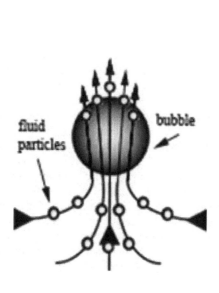

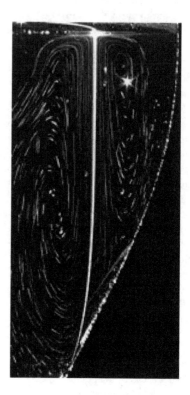

Figure 4. The fluid motion occurs because as the bubbles rise, they drag the fluid along in their wake (left). When seeded with tiny polymer particles and imaged in a time-lapse photo with a laser, the bubble stream appears as a white line, and the regular ring vortex of movement induced in the fluid from the bubble movement is clearly outlined by the particles (right).

We seeded the Champagne with Rilsan particles as tracers of fluid motion. These polymer particles are quasi-spherical in shape, with diameters ranging from 75 to 150 micrometers, and have a density (1.060) close to that of Champagne (0.998). The particles are neutrally buoyant and do not affect bubble production, but they are very reflective of laser light. It is amazing to see the amount of fluid that can be set in motion by viscous effects. In our result- ing images, a central line corresponds to the bubble train path during the exposure time of the camera, and the fluid motion is characterized by a swirling vortex that is symmetrical on both sides of the bubble chain (Figure 6). The vortex-pair in the planar view of our image can be extrapolated to show a three-dimensional annular flow around the center line of bubbles (Figure 6). This means that a single fixed nuclear site on the glass surface can set the entire surrounding fluid into a small-scale ring vortex.

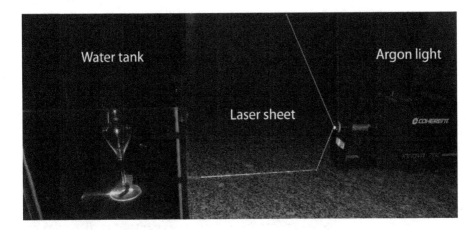

Figure 5. Experimental set-up

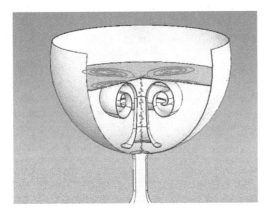

Figure 6. Illustration of the three-dimensional flow in an engraved coupe-glass (right).

5. Controlled mixing

Champagne-tasting science involves a number of very subjective judgments, often difficult to quantify. For example, there is an inherent compromise between the visual aspects of bubbly behavior and olfactory stimulation, as these two qualities appear to be at odds. Too much nucleation will excite the sense of sight but cause the carbonation to quickly fizzle out, making for unpleasant tasting. On the contrary, poor nucleation will produce fewer bubbles in the glass, but more bubbles and aromas in the taster's nose and mouth, consequently enhancing the senses of smell and taste at the expense of sight. From the many experiments we have conducted with controlled effervescence, it seems that an ideal number of about 20 nucleation sites best satisfies this dilemma. Our laser visualizations of fluid flow have shown that a flute with an engraved circular crown reaches a steady state of fluid motion about 30 seconds after the glass is poured [2-5, 7-8]. The vortices do not swirl around and change shape, in contrast to those created in unetched glasses. The bubbles are highly reflective, allowing one to clearly observe the formation of a rising gas column along the vertical glass axis from the treated bottom up to the free surface of the beverage. Consequently, the driving force it imparts to the surrounding fluid generates two large counter-rotating vortices in the vertical lighted section (Figure 6-7). These cells are located outside the rising bubbles, close to the wall of the flute. Because this gas column acts like a continuous swirling-motion generator within the glass, the flow structure exhibits a quasi-steady two-dimensional behavior with a geometry that is symmetrical around the center line of the glass. It clearly appears in the case of an engraved flute that the whole domain of the liquid is homogeneously mixed (Figure 7, right). To complete our observations, we also studied the flow in an engraved traditional Champagne coupe, which is much wider but shallower than the flute. As in the flute, the rising CO_2 bubble column causes the main fluid to move inside the coupe. However, two distinctive steady-flow patterns, instead of one, appear in a glass of this shape. Like the flute, the coupe clearly exhibits a single swirling ring, whose cross section appears as two counter- rotating vortices close to the glass axis. What strongly differs from the motion in the flute is that this recirculation flow region does not occupy the whole volume of the glass. The periphery of the coupe is instead characterized by a zone of no motion. Thus, for a wide-rimmed glass, only about half of the liquid bulk participates in the Champagne mixing process. Nevertheless, in an engraved glass of either shape, the presence of a ring vortex is not time-dependent; it still forms in the coupe, despite the ascent time being about a third of that in the flute.

6. Infrared imaging technique used to visualize the flow of gaseous CO_2 desorbing from champagne

A visualization technique based on the Infrared (IR) thermography principle has been used to film the gaseous CO_2 fluxes outgassing from champagne (invisible in the visible light spectrum) [10]. The CO_2 absorptions observable by the IR camera are quite weak because this gas molecule has only a strong absorption peak in the detector bandwidth at 4.245 mm. Consequently, the best way to visualize the flow of gaseous CO_2 desorbing from champagne is to fit the IR video camera with a band-pass filter (centered on the CO_2 emission peak). The experimental device consists of a CEDIP middle waves Titanium HD560M IR video camera, coupled with a CO_2 filter (Ø 50.8 mm X 1 mm thick– Laser Components SAS). In complement, the technique involves an extended high-emissivity (0.97) blackbody (CI systems provided by POLYTEC PI), used at a controlled uniform temperature of 80°C, and placed approximately 30 cm behind the glass. The IR video camera was used at a 10 frames per second (fps) filming rate.

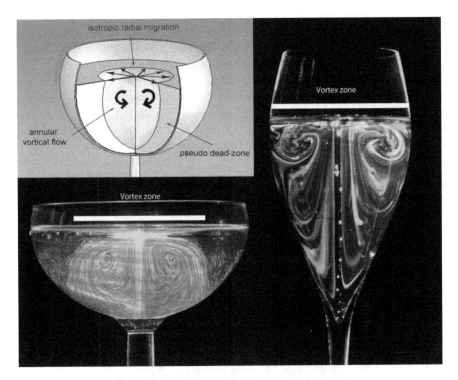

Figure 7. Glass shape and size have great influence on fluid flow and mixing in Champagne and sparkling wines. A flute imaged with fluorescent dye *(right)* shows that the resulting fluid vortex spans the entire width of the glass. A coupe glass, much shorter and wider, imaged with a laser and polymer particles, produces a similar vortex, but the vortex zone only extends across about half of the liquid *(bottom left)*. A dead zone of no motion arises in the outer perimeter of the glass, and bubbles do not reach this area before bursting. A pseudo-dead zone beneath the liquid surface experiences only minimal movement and mixing *(top left)*.

7. Results and discussion

7.1. Losses of dissolved CO2 during the service of champagne in each type of drinking vessel

As recently shown in a previous article, the pouring process is far from being inconsequential with regard to the concentration of CO_2 dissolved into the wine [11]. During the several seconds of the pouring process, champagne undergoes highly turbulent and swirling flows. During this phase, champagne loses a very significant part of its initial content in dissolved CO_2. Gray scale infrared thermography time-sequences displayed in Figure 9 illustrate the progressive losses of dissolved CO_2 desorbing from the liquid phase into the form of a cloud of gaseous CO_2, whether champagne is poured in a flute or in a coupe. Clouds of gaseous CO_2 escaping from the liquid phase clearly appear. Consequently, at the beginning of the time series (i.e., at t=0, after the glass was poured with champagne and manually placed below the sampling valve of the chromatograph), champagne holds a level of dissolved CO_2 well below 11.6 ± 0.3 g L^{-1} (as chemically measured inside a bottle, after uncorking, but before pouring). In the present work, the initial bulk concentration of dissolved CO_2 after pouring, denoted c_i, was also chemically accessed by using carbonic anhydrase. To enable a statistical treatment, six successive CO_2 dissolved measurements were systematically done for each type of drinking vessel, after six successive pouring (from six distinct bottles). When served at 20°C, champagne was found to initially hold (at t=0, after pouring) a concentration of CO_2 dissolved molecules of $c_i^{flute} = 7.4 \pm 0.4$ g L^{-1} in the flute, and $c_i^{coupe} = 7.4 \pm 0.5$ g L^{-1} in the coupe (i.e., approximately 4 g L^{-1} less in both types of drinking vessel after pouring than inside the bottle, before pouring).

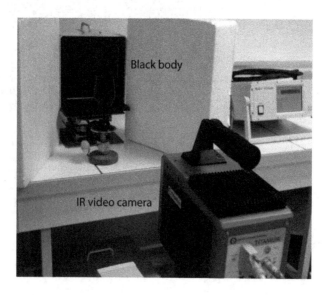

Figure 8. Experimental device

A

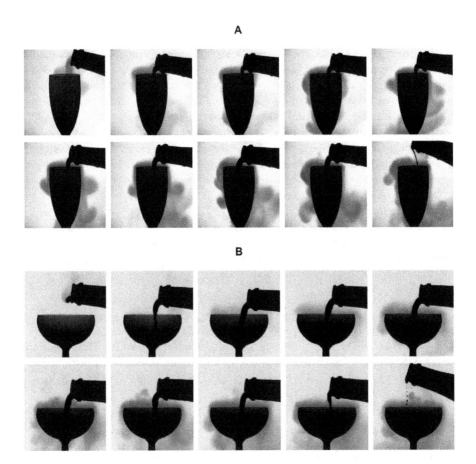

B

Figure 9. Infrared imaging of gaseous CO_2 desorbing when pouring champagne into both glass types. Gray scale time-sequences illustrating the pouring step as seen through the objective of the IR video camera – for a bottle stored at 20°C – whether champagne is served into the flute (a) or into the coupe (b).

7.2. Gaseous CO_2 and ethanol content found in the headspace above each type of drinking vessel

All along the first 15 minutes following pouring, concentrations of gaseous CO_2 found close to the edge of the flute are approximately between two and three times higher than those reached above the coupe. This observation is self-consistent with some recent data about volume fluxes of gaseous CO_2 measurements above glasses poured with champagne, including a flute and a coupe [11]. Fluxes of gaseous CO_2 per unit surface area offered to gas discharging are indeed significantly higher above the surface of the flute than above the surface of the coupe because the same total amount of dissolved CO_2 (<0.7 gram for both glass types after pouring) has to be released by bubbles from a narrower surface, thus concentrating in turn more gaseous CO_2 in the headspace above the flute. Actually, due to higher concentrations of gaseous CO_2 above the flute than above the coupe, the smell of champagne, and especially its first nose, is always more irritating when champagne is served into a flute. It is indeed well-known that a sudden and abundant quantity of CO_2 (a strong trigeminal stimulus) may irritate the nose during the evaluation of aromas [12]. By using time-sequences provided through infrared imaging, the gaseous CO_2 desorbing from champagne and progressively invading the headspace above glasses was made visible in a false color scale (see Figure 10). Such an image processing analysis provides a better visualization of the relative differences in the CO_2 concentration field between both glass types, as shown in the thermography images displayed in Figure 11. Zones highly concentrated in gaseous CO_2 appear in black and dark blue, whereas zones slightly concentrated in gaseous CO_2 appear in red. The concentration of CO_2 found above the flute (close to the edge) is indeed always significantly higher than that found above the coupe. It can be noted for example, through infrared imaging, that the headspace (above the champagne surface, but below the glass edge) remains black during the first 3 min following pouring in case of the flute, whereas it progressively turns blue in case of the coupe.

Moreover, it is also worth noting from infrared imaging time-sequences that the cloud of gaseous CO_2 escaping from champagne tends to stagnate above the glass, or even tends to flow down from the edge of glasses by "licking" the glass walls (rather than diffuse isotropically around them). These observations conducted through infrared imaging betray the fact that gaseous CO_2 is approximately 1.5 times denser ($\rho\ CO2 \approx 1.8\ g\ L^{-1}\ at\ 20°C$) than dry air is ($\rho\ air \approx 1.2\ g\ L^{-1}\ at\ 20°C$), and therefore tends to naturally flow down.

7.3. Numerical modeling of bubble induced flow patterns in champagne glasses

A numerical modeling of flow dynamics induced by the effervescence in a glass of champagne has been carried out for the first time using the finite volume method by CFD (Computational Fluid Dynamics). In order to define source terms for flow regime and to reproduce accurately the nucleation process at the origin of effervescence, specific subroutines for the gaseous phase have been added to the main numerical model. These subroutines allow the modeling of bubbles behavior based on semi-empirical formulas relating to bubble diameter and velocity or mass transfer evolutions. So, the idea of this study is to develop a "universal" numerical modeling allowing the study of bubble-induced flow patterns due to effervescence, whatever the shape of the glass in order to quantify the role of the glass geometry on the mixing flow phenomena and induced aromas exhalation process. Details and development of the steps of modeling are presented in this paper, showing a good agreement between the results obtained by CFD simulations in a reference case of those from laser tomography and Particle Image Velocimetry experiments, validating the present model.

A

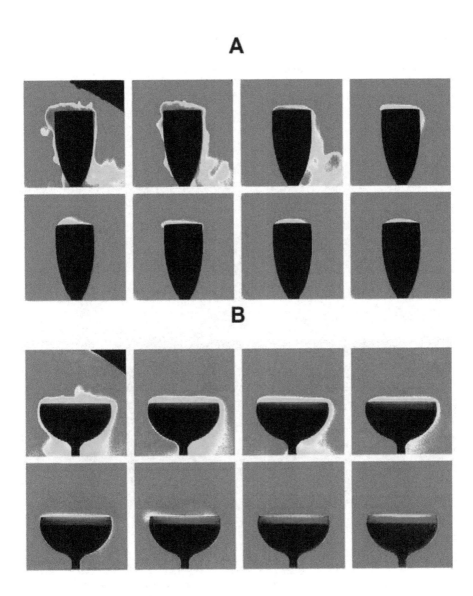

B

Figure 10. Infrared imaging of gaseous CO_2 desorbing from glasses filled with champagne. False color time-sequences illustrating champagne glasses as seen through the objective of the IR video camera, after the pouring step – for a bottle stored at 20°C –whether champagne is served into the flute (a) or into the coupe (b).Zones highly concentrated in gaseous CO_2 appear in black and dark blue, whereas zones slowly concentrated in gaseous CO_2 appear in red.

8. Geometry, equations and boundary conditions

8.1. Geometry and mesh generation

A traditional flute has been considered as a reference glass case. The glass geometry used in this study has been created from the real dimensions measured of the glass used for the experiments.

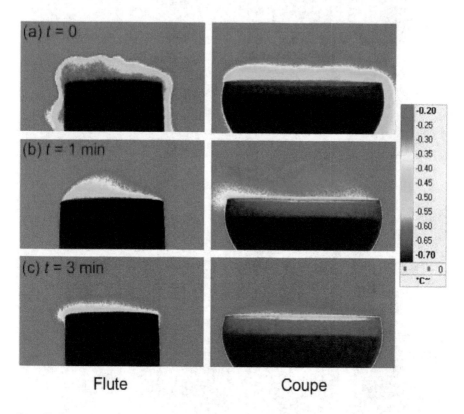

Figure 11. Close-up on gaseous CO_2 desorbing above both glass types. False color IR time-sequences showing close-up snapshots of CO_2 clouds desorbing above the flute and the coupe, respectively, immediately after pouring (a), 1 min after pouring (b), and 3 minutes after pouring (c); By using the color scale which provides a correspondence between the relative abundance of gaseous CO_2 and the temperature detected by the IR sensor of the camera after absorption by the gaseous headspace above glasses, it clearly appears that gaseous CO_2 is always more concentrated above the flute than above the coupe.

To ensure a continuous and perfectly controled process of effervescence, glassmakers usually consider circulary engraved glasses (figure 3). In such a way, as previously mentionned, the flow structure exhibits a quasi steady two dimensional behavior [2] (figures 6-7).

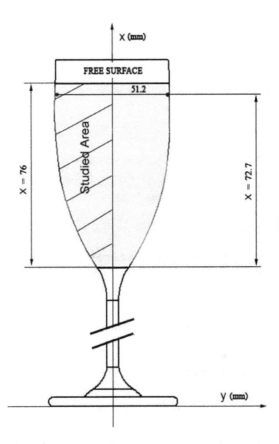

Figure 12. Champagne flute model.

In this situation, a 2D examination in the axis-symmetry plane can be considered as sufficient. For this purpose, only half of the studied area was drawn. The study area has a total height of 76 mm which corresponds to the fill level and its diameter is 51.2 mm at the liquid surface level (figure 12).

The ANSYS® Workbench Design Modeler software has been used to draw the geometry from the real size (scale 1:1) of the numerical study glass used as reference.

The mesh of the domain has been carried out using the ANSYS® Workbench meshing software. It consists in a two-dimensional mesh efficient in the case of simulations of axisymmetrical flow features. The body has been meshed with quadrilateral elements along the central part of the flow curvature, he is structured composed by square elements (L X L) but to follow the wall, the mesh is unstructured (figure 4).

8.2. Equations and numerical scheme

Because champagne is a wine in which as well gaseous as liquid phase are simultaneous present, the flows in a glass of champagne have been simulated numerically with a multi-phase model. The flow is supposed laminar [6,14] and governed by the volume finite equations. The liquid phase hydrodynamics are described with the continuity and momentum conservation equations for laminar flows:

Continuity equation:

The general form of the continuity equation can be written as:

$$\frac{\partial \rho}{\partial t} + \nabla \cdot (\varrho \vec{v}) = S_m \tag{1}$$

The source S_m is the mass added to the continuous phase from the dispersed second phase.

For 2D axisymmetric geometries, the continuity equation is given by:

$$\frac{\partial \rho}{\partial t} + \frac{\partial}{\partial x}(\rho v_x) + \frac{\partial}{\partial r}(\rho v_r) + \frac{\rho v_r}{r} = S_m \tag{2}$$

where x is the axial coordinate, r is the radial coordinate, v_x is the axial velocity and v_r is the radial velocity.

Conservation of momentum is described by:

$$\frac{\partial}{\partial t}(\varrho \vec{v}) + \nabla \cdot (\varrho \vec{v} \vec{v}) = -\nabla p + \rho \vec{g} + \vec{F} \tag{3}$$

where ϱ is the static pressure and $\rho \vec{g}$ and $\vec{F}$ are the gravitational body force and external body forces (forces that arise from interaction between the liquid phase and the dispersed one).

For 2D axisymmetric geometries, the axial and radial momentum conservation equations are given by:

$$\frac{\partial}{\partial t}(\rho v_x) + \frac{1}{r}\frac{\partial}{\partial x}(r \rho v_x v_x) + \frac{1}{r}\frac{\partial}{\partial r}(r \rho v_r v_x) = -\frac{\partial p}{\partial x} + F_x \tag{4}$$

and

$$\frac{\partial}{\partial t}(\rho v_r) + \frac{1}{r}\frac{\partial}{\partial x}(r \rho v_x v_r) + \frac{1}{r}\frac{\partial}{\partial r}(r \rho v_r v_r) = -\frac{\partial p}{\partial r} + F_r \tag{5}$$

where

$$\nabla \bullet \vec{v} = \frac{\partial v_x}{\partial x} + \frac{\partial v_r}{\partial r} + \frac{v_r}{r} \tag{6}$$

In this work, we hve used the Lagrangian-Eulerian approach which analyzes the liquid phase (primary phase) by the Eulerian method and the bubble phase (secondary phase) by Lagrangian assumption allowing the monitoring of bubbles life cycle.

The Euler-Lagrange approach is the basis of the lagrangian discrete phase model. The dispersed phase is solved by tracking the bubbles through the calculated flow domain while the fluid phase is treated continuously by solving the Navier-Stockes equations. Exchanges of momentum and mass are realized between the dispersed phase and the fluid one.

According to the lagrangian multiphase model, the volume fraction of the discrete phase (secondary phase) is quite small.

The bubbles trajectories are computed individually at each time step during the fluid phase calculation. This model is perfectly adapted for modeling the flows in a glass of champagne. In order to reproduce as closely as possible the principle of nucleation, subroutines have been used for the gaseous phase. These subroutines have been written based on physical laws that are taken from experimental results [5, 6].

The trajectory of a bubble is predicted by integrating the force balance in a Lagrangian reference frame.

During its rise in the liquid, a bubble is subjected to the action of several forces [14]:

The buoyancy:

$$F_B = \frac{4}{3}\pi R^3 \rho g \tag{7}$$

The drag force F_D which is related to the fluid flow around the bubble, when the bubble begins to move the fluid that is around it. The movement of the surrounding fluid leads to an additional force F_{MA} called "added mass" related to the variation in the amount of movement of liquid displaced:

$$F_{MA} = \frac{\rho d}{dt}(Vu) \tag{8}$$

The volume V of liquid entrained in the wake of the bubble is roughly equal to the half of the volume of the bubble. Thus:

$$F_{MA} = \frac{2}{3}\rho\pi\frac{d}{dt}\left(R^3 u\right) \tag{9}$$

The equation of motion can be written:

$$\frac{2}{3}\rho\pi R^3\left(\frac{dU}{dt} + \frac{3U}{R}\frac{dR}{dt}\right) = \frac{4}{3}\pi R^3\rho g - \frac{1}{2}C_D\rho U^2\pi R^2 \tag{10}$$

The force of added mass has been compared to the buoyancy along the path of the bubble to the surface. The force of added mass does not exceed 2-3% of the buoyancy, so it can be neglected in the remainder of the study. The equation of motion is finally reduced to a simple equality between the drag force and buoyancy.

In this case, the drag force F_D is defined as

$$F_D = \frac{1}{2}C_D\rho u^2\pi R^2 \tag{11}$$

where C_D is the drag coefficient. During ascent, surface active materials progressively accumulate at the rear part of the rising bubble, thus increasing the immobile area of the bubble surface.

Arising bubble rigidified by surfactants runs into more resistance than a bubble presenting a more flexible interface free from surface-active materials. The champagne bubbles showed therefore a behavior intermediate between that of a rigid and that a fluid sphere. To take into account the surfactants accumulation, the two following experimental drag coefficients laws, available in the range of intermediate Reynolds numbers (10^{-1} to 10^2) covered by champagne bubbles, have been used.

Magnaudet et al [14] have proposed a semi empirical relationship between the drag coefficient and the Reynolds number:

$$C_D = \frac{16}{Re}\left(1 + 0.15\sqrt{Re}\right) \qquad (Re < 50) \tag{12}$$

This experimental determination of the drag coefficient for fluid spheres is available for Reynolds number less than 50.

Since the Reynolds number exceeds the limit of 50 for sufficiently long path, another empirical law has been used available for $Re > 50$, determined by *Maxworthy et al* [14]

$$C_D = 11.1Re^{-0.74} \qquad (1 < Re < 800) \tag{13}$$

Because bubbles do not exceed a critical diameter of 2 mm, they are spherical during their ascent. Moreover, the assumption that the bubbles do not coalesce or breakup has been considered.

The Reynolds number is defined by:

$$R_e = \frac{2\rho R u}{\eta} \tag{14}$$

The density and the viscosity of the champagne wine for the liquid phase and the density and the viscosity of the carbon dioxide for the gaseous phase have been stored in the materials database (table 1).

The numerical simulations have been carried out with the ANSYS FLUENT® software using volume finite approach. The convergence criteria were based on the residuals resulting from the integration of the conservation equations over finite control-volumes. During the iterative calculation process, these residuals were constantly monitored and carefully scrutinized. For all simulations performed in this study, converged solutions were usually achieved with residuals as low as 10^{-5} (or less) for all the governing equations. To carry out numerical simulations on the dynamics of the fluid, a structured mesh, whose dimensions are 0.2 x 0.2 mm², has been chosen in the main central part of the domain where bubbles are present.

9. Boundary conditions

Models based on classical nucleation theory do not give a satisfactory approach of nucleation in the effervescent wines [13]. The idea has been to create routines to simulate the principle of nucleation and then compare the results with those obtained with experimental data [5, 6]. In order to simulate the bubbles growth, the bubbles velocity, the mass transfer between bubbles, the mass flow rate and the drag law, we have written User Defined Functions (UDF) in C language which defines source terms for the flow regime. The variation of bubbling frequency, which is a function of the CO_2 dissolved concentration, is made possible by changing the time step during the calculation.

The radius R(m) of champagne bubbles increase in time at a constant growth rate $k = \frac{dR}{dt}$, as bubbles rise toward the liquid surface. Thus, $R(t) = R_0 + kt$ where R_0 is the bubble radius as it detaches from the nucleation site.

The semi-empirical growth rate (k, μm/s) of bubbles rising in champagne was linked with some physicochemical properties of liquids as follows [5, 6]:

$$k = \frac{dR}{dt} \approx 0.63 \frac{R_0}{P_0} D_0^{\frac{2}{3}} (2\alpha\rho g / 9\eta)^{1/3} (C_L - k_H P_0)$$

(15)

The bubbles diameter depends on both the distance to the surface H (m) and the growth rate k (μm/s) which decreases over time. The law governing the change in radius R(m) of a bubble is [5, 6]:

$$R \approx 3\left(\frac{\eta}{2\alpha\rho g} kH\right)^{1/3}$$

(16)

Where α is a numerical coefficient that depends on the fluid in question, estimated to be 0.7 in the case of sparkling wines [5, 6].

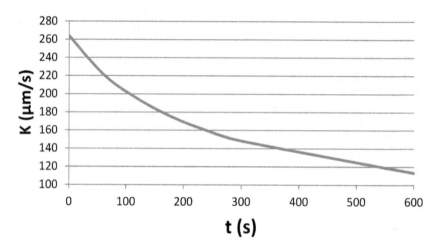

Figure 13. Evolution of the bubble growth rate according to time after pouring.

The bubbles velocity u(m/s) varies according to the following expression [6, 14]:

$$u \approx \frac{2\alpha\rho g}{9\eta} R^2 \qquad (17)$$

This factor accounts the rigidity of the bubble and thus the braking effect due to the presence of surfactant molecules on the surface of bubbles. It will be smaller if the bubble is made rigid by a thick shield of surfactant molecules.

The mass flow rate Q_m (kg/s) is defined by:

$$Q_m = \frac{N}{t} \times \rho_p \times V_b \qquad (18)$$

The CO_2 dissolved concentration C_L decreases continuously over time once the wine is poured into the glass. This parameter also varies with the temperature. In this study, we have used the champagne physicochemical parameters at 20° C corresponding to our reference temperature for the experiments (**Table 1**).

	Champagne (20°C)	Carbon dioxide (20°C)
Density (kg/m³)	998	1.7878
viscosity (kg/m/s)	0.00166	$1.37 \cdot 10^{-5}$
Surface tension (N/m)	0.0468	0

Table 1. Physicochemical parameters of Champagne and Carbon dioxide (from [14]).

10. Numerical results

The results obtained by numerical simulation have been compared with those from experiments using two flow visualization techniques to get both qualitative and quantitative viewpoints (figure 14) [2-5]. The laser tomography as a qualitative analysis method has been used to visualize the flow patterns and vertical structures induced by the continuous column of ascending bubbles in the reference flute poured with champagne. A comparison of the flow feature is presented in Figure 9. During the time-exposure of a camera, the liquid seeded with solid Rilsan particles [2-5, 7-8] and lighted by a planar laser sheet exhibits streamline patterns (figure 14b). Comparison between these experimental streamlines and the numerical ones (figure 14a) shows a good agreement, especially regarding to the location of the vortex cores in the investigated domain. The global flow features are satisfactory modeled with the CFD developed code.

To highlight a quantitative validation of the numerical modeling, velocity profiles as well velocity profiles as velocity iso-contour maps have been interested. The experimental data have been obtained by Particle Image Velocimetry measurements [15]. Figure 15 presents a comparison between the two experimental and CFD velocity profiles drawn for a 72.7mm X-location. The general trend is well reproduced as well for the X-velocity peak on the axis of symmetry as for the return flow characterized by negative velocity X-component values. Curve extrema are located at the same Y-location. Numerical results are in good accordance with those obtained by PIV measurement for the velocity profiles. Similar conclusions have been deduced for other X-locations (not presented here).

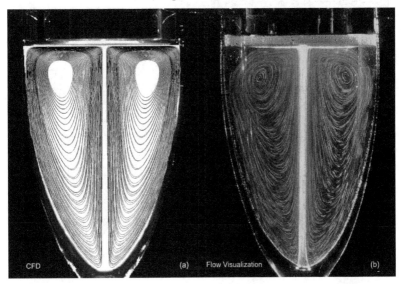

Figure 14. Streamlines obtained by CFD simulation (a) compared with classic flow visualization (b) at t = 5 minutes following the pouring process.

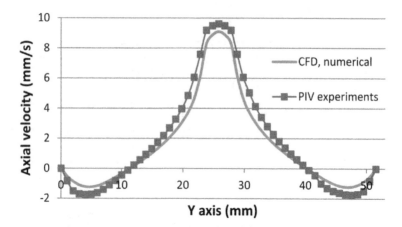

Figure 15. Axial velocity of liquid phase at t = 1 minute after pouring process and X = 72.7 mm, comparison between PIV measurement and CFD simulation.

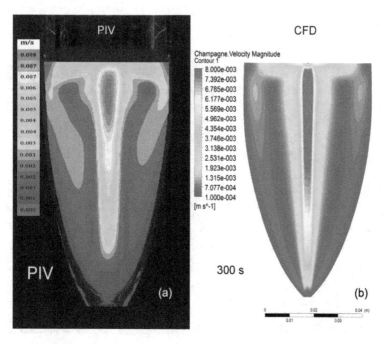

Figure 16. PIV measurements (a) compared with CFD simulation (b) at t = 5 minutes after pouring process, velocity magnitude of the liquid phase.

Velocity-magnitude maps are drawn in figure 16 for the two numerical and PIV-measurement approaches. A close agreement appears on the two maps. A same maximum velocity magnitude whose value is 8 mm/s is observed at the same location on the central part of the flow. Similar comments can be drawn concerning the vortex cores locations. Even if iso-contour curves differ a little bit, the trend is duplicated on these two maps and the estimated velocity is the same order of magnitude as those measured by experimental one.

Thus, one can conclude that the numerical simulation allows a satisfactory approach of the fluid dynamics.

11. Conclusion

A classical flow visualization technique was used in order to capture the fluid motion in traditional flutes and coupes poured with champagne. It was found that glasses engraved around their axis of symmetry produce a rising gas column along the vertical glass axis which induces, in turn, recirculating flow regions. In case of the classical engraved champagne flute, the whole domain of the liquid phase is homogeneously mixed, whereas in the case of the engraved champagne coupe, the recirculating flow region does not occupy the whole volume in the glass. In the engraved coupe, a "dead-zone" of no motion was identified which inhibits the formation of the collar at the glass edge. Because the kinetics of flavor and gas release also strongly depend on the velocity of the recirculating flows close to the interface, we therefore strongly believe that this paper brings objective elements and clues in order to better understand the role of glass shape and engravement conditions on the "olfactive" behavior of champagne and sparkling wines in a glass. To go further; a developed gaseous CO_2 visualization technique based on infrared imaging was performed. Those analytical results are self-consistent with sensory analysis of champagne and sparkling wines, since it is generally accepted that the smell of champagne, and especially its first nose, is always more irritating (because more concentrated in gaseous CO_2 which is a strong trigeminal stimulus) when champagne is served into a flute than when it is served into a coupe. In addition, a numerical modeling of flow dynamics induced by effervescence in a glass of champagne has been carried out for the first time in order to quantify the role of the glass geometry on the mixing flow phenomena and induced aromas exhalation process.

Appendix

C_D drag coefficient (dimensionless)

C_l lift coefficient (dimensionless)

C_L CO_2 dissolved concentration in the liquid (g/l)

D_0 diffusion coefficient of CO_2 molecules (m² / s)

F_D drag force

g acceleration due to gravity (m.s^{-2})

H liquid height (m)

k theoretical growth rate of bubbles (μm/s)

L characteristic dimension of a mesh element (mm)

N bubbling frequency (H_z)

P_0 atmospheric pressure (atm)

Q_m mass flow rate (kg/s)

R bubble radius (m)

R_e Reynolds number (dimensionless)

u bubble velocity (m/s)

v liquid velocity (m/s)

V_b bubble volume (m^3)

R ideal gas constant (8.31 J/mol/K)

ρ liquid density (kg/m^3)

ρ_p density of CO2 (kg/m^3)

η dynamic viscosity (kg/m/s)

α numerical coefficient (dimensionless)

θ liquid temperature (°C)

λ molecular mean free path (m)

ϕ bubble diameter (m)

Author details

Fabien Beaumont[1], Gérard Liger-Belair[2] and Guillaume Polidori[1*]

1 GRESPI/Thermomécanique, Université de Reims, France

2 GSMA, UMR CNRS 7331, Université de Reims, France

References

[1] Liger Belair G., Cilindre C., Gougeon R.D., Lucio M., Gebefügi I., Jeandet P., Schmitt-Kopplin P., Unraveling different chemical fingerprints between a champagne wine and its aerosols, PNAS 2009, volume 106, n°39, 16545-16549.

[2] Polidori G., Jeandet P., Liger-Belair G., Bubbles and flow patterns in Champagne, American Scientist 2009, 97, 294.

[3] Liger Belair G., Religieux J.-B., Fohanno S., Vialatte M.-A., Jeandet P., Polidori G., Visualization of mixing phenomena in champagne glasses under various glass-shape and engravement conditions, J. Agric. Food Chem 2007, 55, 882.

[4] Polidori G., Beaumont F., Jeandet P., Liger Belair G., Artificial bubble nucleation in engraved champagne glasses, J. Visualization 2008, 11-4, 279.

[5] Liger Belair G., Polidori G., Jeandet P., Recent advances in the science of champagne bubbles, Chem. Soc. Rev. 2008, 37, 2490.

[6] Liger Belair G., Ann.Phys 2002. (Paris) 27, 1.

[7] Polidori G., Beaumont F., Jeandet P., Liger Belair G., Ring vortex scenario in engraved Champagne glasses, J. Visualization 2009, 12-3, 275.

[8] Polidori G., Beaumont F., Jeandet P., Liger Belair G., Visualization of swirling flows in champagne glasses, J. Visualization 2008, 11-3, 184.

[9] Liger Belair G., Beaumont F., Jeandet P.and Polidori G., Flow patterns of bubble nucleation sites (called fliers) freely floating in champagne glasses, Langmuir 2007, 23, 10976.

[10] Liger-Belair G, Bourget M, Villaume S, Jeandet P, Pron H, et al., On the losses of dissolved CO_2 during champagne serving, J Agric Food Chem 2010, 58: 8768–8775.

[11] Liger-Belair, G., Villaume, S., Cilindre, C., Jeandet, P., CO_2 volume fluxes outgassing from champagne glasses: The impact of champagne ageing, Analytica Chimica Acta 2010, 660, 29–34.

[12] Duteurtre B, Le Champagne: de la tradition à la science 2010. Paris: Lavoisier. 384 p.

[13] Herrmann E., Lihavainen H., Hyvärinen A.P., Riipinen I., Wilk M., Stratmann F., Kulmala M., Nucleation Simulations using the fluid dynamics software FLUENT using the Fine Particle Model FPM, The Journal of physical chemistry 2006, A 2006, 110, 12448-12455.

[14] Liger Belair G., Marchal R., Robillard B., Dambrouck T., Maujean A., Vignes-Adler M., Jeandet P., On the Velocity of Expanding Spherical Gas Bubbles Rising in Line in Supersaturated Hydroalcoholic Solutions: Application to Bubble Trains in Carbonated Beverages, Langmuir 2000, 16, 1889-1895.

[15] Liu Z., Zheng Y., Jia L., Zhang Q., Study of bubble induced flow structure using PIV ,
       Chemical Engineering Science 2005, 60, 3537-3552.

Chapter 3

Mass Transfer Over the Surface of Metal Nanostructures Initiated and Stopped by Illumination

T.A. Vartanyan, N.B. Leonov, S.G. Przhibel'skii and
N.A. Toropov

Additional information is available at the end of the chapter

1. Introduction

Metal nanostructures play an increasing role in modern technologies. Being far from the thermodynamic equilibrium they are vulnerable to the influence of external agents. In particular, thermal stability of metal nanostructure is an issue in many applications. In this contribution we are going to discuss one counterintuitive way of stabilization of metal nanostructure, namely, illumination. It is generally agreed that illumination will destroy nanostructure if do not leave it intact. Nevertheless, we found that in some circumstances illumination may lead to tiny restructuring of the nanostructure surface that improves its thermal stability. This restructuring is not readily observable after illumination but lead to a pronounce effect after annealing. Illuminated nanostructures acquire immunity to the heating and change their structure much more slowly than the nanostructures that were not illuminated in advance. Moreover, the final state of the illuminated nanostructures after annealing differs considerable from that of unilluminated one being much more similar to its initial state. Taking into account close connection between the structure and the optical properties of the plasmonic metal nanostructures it is not surprising that after annealing illuminated and non-illuminated regions may be differentiated by eye.

More traditionally, illumination may initiate structural changes of the metal nanostructures. In particular, reshaping of metal nanoparticles that support plasmon excitation was studied by optical means. Resonance enhancement of the transformation rate was observed in our experiments on the arrays of alkali metal nanoparticles.

On the other hand, illumination can initiate structural transformations of the organic molecules adsorbed on the surface of metal nanostructures. In our experiments silver nanoparticles were prepared by vacuum evaporation on a sapphire substrate. Cyanine dye molecules were spread over the silver nanoparticle arrays by spin-coating technique. The samples were characterized by scanning electron microscopy and optical spectroscopy. A significant increase of the dye photoinduced transformation rate was observed. Simultaneously, an enhanced absorption and fluorescence were observed.

The extinction spectrum of the hybrid material was rationalized as a result of mutual interactions between the plasmon oscillations localized in the metal nanoparticles and resonance absorption and refraction of dye molecules. Plasmon resonances are shifted due to the anomalous refraction of dye molecules. Depending on the spectral position of the dye absorption band relative to the inhomogeneously broadened plasmon band this shift may lead to considerable clarification of the sample at particular wavelengths that was observed experimentally. On the other hand, the absorption of dye molecules is enhanced due to the incident field amplification in the near field of metal nanoparticles. Even when the dye absorption band overlaps with the tail of the plasmon band of silver nanoparticles, 3 to 5 times enhancement of the dye absorption was obtained. Besides that a nearly 4-fold increase of cyanine dyes fluorescence intensity in the presence of metal nanoparticles was observed.

The photoinduced transformations of the dye molecules situated in the near field of the metal nanoparticles were studied. The rate of the transformations on the surface of metal nanoparticles was found to increase as compared to that on the surface of a dielectric substrate.

2. Illumination as a control over the surface mass transfer

Mass transfer over the solid surfaces is of crucial importance in many technological processes like vacuum vapor deposition, heterogeneous catalysis and many others. For example, when the granular metal films are grown via Volmer-Weber growth mode the atoms evaporated from the source first impinge on the surface and then diffuse over it. A nucleus of a metallic phase is formed by several atoms met together after a long way through the surface made separately by each of them. Then, the nucleus grows incorporating new atoms that first diffuse over the substrate and then climb up the existing flat metal cluster to form a 3D nanoparticle. All these processes being random, the resultant nanostructure is highly irregular. Figure 1 plots the optical extinction spectrum of silver granular film grown on a sapphire substrate. The broad band with a maximum at 420 nm shown as curve 1 is due to the absorption by surface plasmons localized in the silver nanoparticles. The width of this band is so large because there are nanoparticles of different shapes in the film. In particular, the red tail of the plasmon band is due to the flat pancake-like nanoparticles. These shapes are metastable and transform into more round shapes when heated. The band shifts in this case to the blue as a whole.

Illumination provides a means of much more delicate action upon the particle. Being irradiated by ruby laser the granular film changes in such a way that its extinction spectrum acquire the form shown in Figure 1 by curve 2. The shape of the extinction spectrum evidenced the light initiated mass transfer process over the surface of metal nanoparticles.

Indeed, one can notice that the number of flat particles that contribute to the red absorption is reduced while the number of the more round particles that contribute to the yellow absorption is increased. Thus, laser illumination may be used to the fine tuning of the shape distribution of metal nanoparticle. This possibility may be employed to obtain the values of the plasmon dephasing times [1,2].

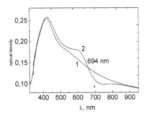

Figure 1. Light-induced mass transfer over the surface of silver nanoparticles. After illumination by three pulses of ruby laser with the fluence of 44 mJ/cm² the original extinction spectrum of the silver granular film on sapphire (1) transforms into (2). The observed changes in the extinction spectrum are due to the thermal diffusion of silver atoms over the surface of metal nanoparticles comprising the granular film.

The observed changes of the extinction spectrum are due to the contraction and reshaping of the nanoparticles in the course of the laser illumination. The sign variation of the optical density difference bears witness of an important role played by surface diffusion. Should evaporation be the sole consequence of the laser heating of the nanoparticles, one expects to observe only a dip centered at the laser frequency. Contrary to that, diffusion leads to the dip accompanied by a bump at a shorter wavelength, the laser frequency being in the midpoint between them. In reality both processes take place. A good fit to the experimental observations may be obtained with the formula that takes into account evaporation as well as diffusion. According to [3] the optical density difference ΔD is a function of the normalized laser detuning

$$x = \frac{\omega - \Omega}{\Gamma},$$ (1)

where Ω is the laser frequency, ω is the frequency at which one observes the extinction difference and Γ is the dephasing rate of the plasmon resonance localized in the nanoparticles. This function reads as

$$\Delta D(x) = -A\frac{1}{1+x^2} + B\frac{x}{\left(1+x^2\right)^2},$$ (2)

where A and B are the constants that define the relative contributions of the evaporation and diffusion processes, correspondingly. Figure 2 plots the optical density difference ΔD obtained experimentally as well as the theoretical curve computed according to Eqs. (1) and (2) with the following parameters $A=0.013$ and $B=0.36$. As B is much larger than A, one may conclude that it is the diffusion of silver atoms over the surface of silver nanoparticle that contributes most in the mechanism of the spectral hole burning while the contribution of evaporation is rather small.

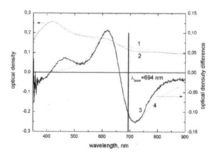

Figure 2. Optical density difference ΔD obtained experimentally (3) and the theoretical fit (4). Both curves are related to the right vertical axis. Two other curves related to the left vertical axis represent the optical density of the film before and after illumination.

To make a meaningful fit a slight shift of the laser frequency is required. As it was shown in [4], when the optical density of the film is considerable, as it is in our case, reflection is to be accounted for even though an individual particle absorb the light much stronger than scatter it. This additional contribution to the optical density of the film shifts the spectrum but does not affect the relative values of the diffusion and evaporation contributions.

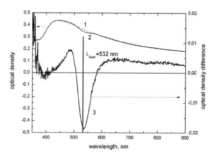

Figure 3. The original extinction spectrum of the silver granular film on sapphire (1) transforms into (2) after illumination by second harmonic of Nd:YAG laser. The optical density difference ΔD before and after illumination is given by curve (3). In this case contribution of the evaporation process is more pronounced than in the case of ruby laser illumination.

Figure 3 plots the hole burnt in the extinction spectrum of the silver granular film after illumination by the second harmonic of the Nd:YAG laser at 532 nm. Obviously, the dip follows the laser frequency. On the other hand, relative contributions of the diffusion and evaporation processes in this case are different. Evaporation is much more evident in this case as compared with the case of ruby laser.

3. Initiation of the surface diffusion via a non-thermal laser action

Mass transfer described in the previous section was initiated by a crude force of a pulsed laser induced heating. More intriguing processes may be induced on the nanoparticle surface illuminated by cw lasers with low intensity. Heating is not operative in this case. Nevertheless, non thermal photo induced diffusion still causes the reshaping of metal nanoparticles. As the spectral position of the plasmon resonance is very sensitive to the particle shape, absorption spectroscopy may be employed to reveal this phenomenon. In the Figure 4 we present the results of the selective action of the low intensity cw diode laser on the ensemble of sodium nanoparticles on sapphire substrate. Due to the high reactivity of sodium this experiment was performed in a sealed glass cell with sapphire windows. Excitation of the surface plasmons localized in the sodium nanoparticles is responsible for the broad bell-shaped curve in the extinction spectrum. Illumination of the sodium granular film by a diode laser operating at the wavelength of 875 nm for 600 s with the power of 40mW produces tiny changes in the extinction spectrum. Although being small these changes are clearly visible in the differential spectrum also plotted in the Figure 4. Figure 5 and 6 plot the results of the similar experiments with lasers operating at the shorter wavelengths 852 nm and 810 nm. A striking feature of this experiments is that the laser action on the sodium nanoparticle shape is selective. Indeed, the zero crossing point of the differential spectrum is close to the laser wavelength and follows it as the laser wavelength changes. This may be clearly seen the Figure 7 where the position of the zero crossing point is plotted against the laser wavelength. At present we do not have a theoretical description of the spectral profile of the differential spectrum in this case. Nevertheless, one sees that general features are similar to the case of the heating induced reshaping. Non-thermally induced diffusion leads to the blue shift of the resonance frequencies of the plasmon resonances. Thus, the particles acquire more round shapes.

4. Inhibition of the surface diffusion by UV irradiation

Thin metal films on dielectric substrate are widely used in different applications ranging from microelectronics to heterogeneous catalysis. In most cases it is desirable that the structure preserves its properties for a long time. In particular, it is of practical importance to improve the thermal stability of the films. On the other hand, to reach such stability is not easy because the thin films are metastable. Thermal stability of the thin silver films is actively studied [5-7] because the agglomeration speeds up at the elevated temperature and leads to the undesirable changes in the film morphology as well as its chemical, electrical and optical properties.

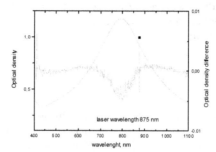

Figure 4. The broad bell-shaped curve represents the absorption band due to the surface plasmons localized in the sodium nanoparticles on the sapphire substrate. Tiny changes of the extinction spectrum after illumination for 600 s with 40mW cw diode laser at 875 nm are revealed by plotting the difference spectrum.

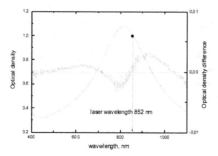

Figure 5. Same is Figure 4 with cw diode laser illumination at 852 nm.

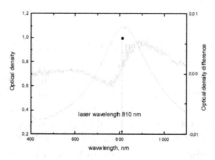

Figure 6. Same is Figure 4 with cw diode laser illumination at 810 nm.

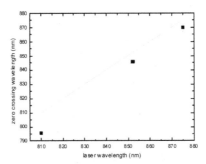

Figure 7. Selectivity of the laser action is made evident by the strong correlation between the position of the zero crossing point (squares) and laser wavelength. The straight light aids the eyes to see that this correspondence is not perfect.

Here we report on the new phenomenon of thin silver film stabilization under UV illumination. The illumination itself does not lead to any noticeable changes in the film appearance. At the same time, the illuminated films demonstrate enhanced stability against heating.

The films were obtained by thermal evaporation of silver on a sapphire substrate kept at the room temperature. The film prepared in vacuum was transferred into a glass cell with the sapphire windows. Then, the cell was evacuated and, finally, sealed off the vacuum system. UV radiation was produced by a mercury lamp. An optical filter was used to deliver radiation in the range from 300 to 400 nm. The intensity of UV illumination was set to 20 mW cm^{-2}. The duration of illumination was typically about 3 hours. To be on the safe side, only one half of the film was illuminated with the UV light. Another part of the film was used as a reference that undergoes just the same annealing procedures as the part, which was treated with the UV light.

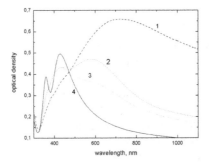

Figure 8. Extinction spectra of the thin silver films on sapphire: Extinction spectra of silver films on sapphire. Curve 1 was obtained after deposition and holding at room temperature for 3 hours. Curves 2 and 4 show the results obtained after annealing of the unirradiated film at 200°C for 5 and 30 minutes, respectively. Curve 3 shows the result obtained after annealing of the irradiated film at the same temperature for 30 minutes.

Figure 8 shows the results obtained in the course of annealing of the films. Just after the deposition the films undergo although very slight but noticeable changes. For this reason, the curve designated as 1 represents the extinction spectrum of the film that was hold at room temperature for 3 hour. This time required for the film to come into its metastable state that may be preserved for many days. Illumination with the UV radiation does not lead to any changes of the extinction spectrum. Hence, curve 1 represents the extinction spectra of the irradiated as well unirradiated film. In the course of annealing at 200°C the unirradated film passes through the states with the extinctions represented by curves 2 and 4. Curve 2 corresponds to annealing for 5 minutes and curve 4 corresponds to annealing for 30 minutes. Curve 3 corresponds to the annealing of the irradiated film for 30 minutes. Obviously, the annealing of the irradiated film proceeds much slowly than that of the unirradiated film. The final result of the annealing of the irradiated film is not so dramatic as that of unirradiated film.

Theses observations obtained via optical means were confirmed by scanning electron microscopy. Microphotographs are shown in Figure 9. After annealing of an unirradiated film its structure transforms into an array of well separated nearly spherical nanoparticles that may be seen in the image 4 of Figure 9. These changes are in accord with the blue shift and narrowing of the plasmon resonance represented by curve 4 in Figure 8. On the other hand, in image 3 of Figure 9 one finds much more in common with the image 1 of the initial film. This corresponds to the optical extinction spectrum 3 with a prolonged red tail resembling the red tail of the initial film 1.

It is to be mentioned that these results are representative in that sense that annealing at higher temperatures (up to 280°C) and for longer times (up to 3 hours) changes nor the morphology neither extinction spectra of the films.

Using of the UV light is essential for obtaining the described stabilization effect. We tried the irradiation at the wavelength of 440, 530 and 810 nm with the same intensity and duration of the irradiation and seen no effect of irradiation at this wavelengths on the films thermal stability.

5. Light induced transformations in the layers of organic molecules spread over the granular metal films

Cyanine dye molecules include a chain made of an odd number of methine groups bound together by alternating single and double bonds and two heteroaromatic rings at both ends of the chain. The structure of the cyanine dye molecules used in our experiments is shown in the Figure 10. Being adsorbed on the surface of a solid material, dye molecules form a layer with several preferable orientations of the long molecular axis relative to the surface normal. Heating and illumination are known to cause the reversible as well as irreversible mass transfer in the molecular layer that may be used for sensing and information recording applications [8,9]. It is tempting to study the possibility of enhancing these properties of the dye molecular layers by spreading them over the metal nanostructures. If the absorption bands of the plasmon excitations in the metal nanostructure coincides or lies close to the absorption bands of the dye molecules one expects the mutual interaction between two resonances that may lead both to their enhancement as well as inhibition.

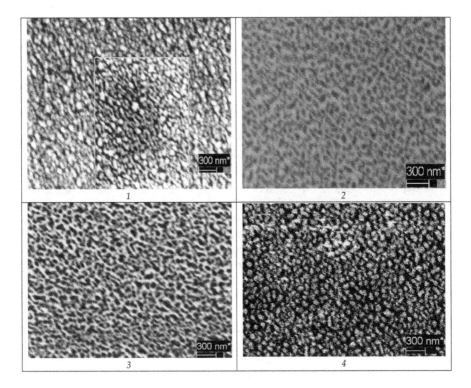

Figure 9. SEM images of silver films: Micrograph *1* represents the original unirradiated film. Micrographs *2* and *4* represent the results obtained after annealing of the unirradiated film at 200°C for 5 and 30 minutes, respectively. Micrograph 3 represents the result obtained after annealing of the irradiated film at the same temperature for 30 minutes.

Figure 10. Structure of the cyanine dye molecule used in the experiments.

In our experiments silver nanoparticles were prepared by vacuum evaporation on a sapphire substrate. Cyanine dye molecules were spread over the silver nanoparticle arrays by spin-coating technique. To achieve the homogeneity of the molecular layers the substrate spins with the rate of 4000 revolutions per minute. The dye layer thickness was set by the concentration of the solution used for spin coating. It was further checked by dissolution of the dye layer in the known quantity of solvent and determination of the dye concentration in it via optical means. The dye layer thickness in our experiments varied from 0.1 to 10 monolayers.

A special attention was paid to the stability of the granular metal film in the course of the dye deposition by spin coating technique. It was found that the as prepared granular metal film is not stable enough to ensure the reproducibility of the results. For this reason the granular metal films obtained via vacuum vapor deposition were suspended in ethanol overnight. After this treatment the optical density of the granular metal films reduces substantially due to the loss of the metal particles with low adhesion to the substrate. Those particles that remain on the surface after such treatment withstand many cycles of spin coating by dyes and subsequent rinsing in ethanol without noticeable changes in their morphology and optical properties.

Figure 11 plots the extinction spectrum of the dye molecular layer coated on the granular silver film on the sapphire substrate (curve 3). For comparison, extinction spectra of the dye molecular coated on the bare sapphire substrate (curve 1) and the granular silver film without dye (curve 2) are shown as well. The dye molecules contribution to the extinction of a hybrid material is revealed by the curve (4) that is the difference between (3) and (2). Clearly, the dye molecules absorbed much stronger when they are put in the vicinity of the silver nanoparticles as compared to the dye molecules on the bare sapphire substrate, the surface densities of the dye molecules being the same in both cases.

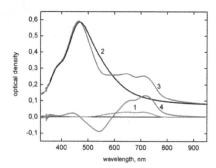

Figure 11. Extinction spectra of the dye layer on the sapphire substrate (1), the granular silver film on the sapphire substrate (2), and the dye layer coated over the granular silver film on the sapphire substrate. The curve (4) plots the difference between (3) and (2), i.e. the dye molecules contribution to the extinction of a hybrid material. Clearly, the dye molecules absorbed much stronger when they are put in the vicinity of the silver nanoparticles as compared to the dye molecules on the bare sapphire substrate, the surface densities of the dye molecules being the same in both cases.

The curve (4) in the Figure (11) demonstrates not only the fact that the absorption by dye molecules is enhanced in the vicinity of silver nanoparticles but also a more subtle effect of the reduced absorption of silver nanoparticles being coated by dye molecules. This may be rationalized by considering the inhomogeneous broadening of the plasmon band and the anomalous dispersion associated with the absorption bands of the dye molecules. It is well known that the spectral position of the plasmon resonance depends on the refractive index of the surrounding material. Near the blue edge of the dye absorption band the anomalous dispersion leads to the low values of refractive index. Hence, the frequency of the plasmon resonance is expected to rise and its spectral position to shift in the blue direction. As in the spectral range under consideration the relative concentration of the nanoparticles rises with the rise of the resonance frequency, the above mentioned sift of the resonance frequencies of plasmon oscillations leads to the observed reduction of the hybrid material extinction. One can expect that the extinction maximum shifts as well. Such a shift indeed observed but it is rather small due to the damping of the plasmon resonance by interband transitions in silver.

The photoinduced transformations of the dye layers were observed under the action of the of the second harmonic of the Nd:YAG laser at the wavelength of 532 nm. Five pulses of 8 ns duration were delivered on the surface. The fluence was kept at the value of 8 mJ cm^{-2} to avoid burning of the spectral holes described above. First, the laser induced transformations were observed in the molecular layers on the bare sapphire substrate. Figure 12 plots the optical density of the dye layer before illumination (1) and after illumination (2). These changes may be interpreted as the departure of the molecules that form the aggregates from each other. In the extinction spectrum this process is seen as the reduction of absorption in the wings and the increase of absorption in the central part the band.

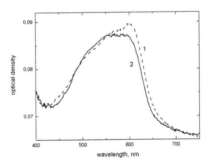

Figure 12. Optical density of the dye layer on the bare sapphire substrate before illumination (1) and after illumination (2).

The molecular movement under illumination in the case of the dye layer coated on the granular metal film is quite different. Figure 13 plots the optical density of the dye layer before illumination (1) and after illumination (2). To facilitate comparison with the case of the bare sapphire substrate absorption of the silver granular film was subtracted. One can easily see that even the sign of the laser induced is different from the case of dye layer on the bare sapphire substrate. As absorption in the wings rises one can conclude that the dye molecules moves to each other to produce the aggregated forms.

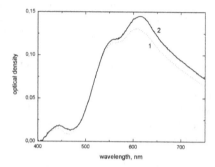

Figure 13. Optical density of the dye layer coated on the granular silver film on the sapphire substrate before illumination (1) and after illumination (2).

6. Conclusion

In this contribution we have presented experimental evidences that illumination is a convenient control of surface mass transport. This starts with the ordinary thermal action of light. In the case of granular metal films that consist of a collection of nanoparticles, this thermal action may be highly selective because the narrow band laser irradiation interacts strongly only with those particles that posses the plasmon resonance at the laser wavelength. Thus, although the diffraction limit avoids possibility to restrict the area of interaction tighter than a fraction of the wavelength, selectivity of the laser action is obtained in the spectral domain. Even the low intensity illumination may be used to initiate the mass transfer over the surface of metal nanoparticles. The selectivity of the laser action has been demonstrated in this case too.

An important issue of the thermal stability of thin metal films may be addressed with the ultraviolet irradiation. In this case we have shown that the granular silver film irradiated with UV light reduces its susceptibility to the thermal stress to a great extent.

Finally it was shown that the dye molecular movement initiated by illumination is quite different in the case of the dye molecules spread over the bare sapphire substrate and the granular metal film on the sapphire substrate.

Acknowledgments

This work was supported in part by Russian Foundation for Basic Research under grant #11-02-01020.

Author details

T.A. Vartanyan*, N.B. Leonov, S.G. Przhibel'skii and N.A. Toropov

St. Petersburg National Research University of Information Technology, Mechanics and Optics, St. Petersburg, Russian Federation

References

[1] Stietz, F., Bosbach, J., Wenzel, T., Vartanyan, T., Goldmann, A., & Träger, F. Decay Times of Surface Plasmon Excitation in Metal Nanoparticles by Persistent Spectral Hole Burning. Phys. Rev. Lett. (2000). , 84(24), 5644-5647.

[2] Bosbach, J., Hendrich, C., Stietz, F., Vartanyan, T., & Träger, F. Ultrafast dephasing of surface plasmon excitation in silver nanoparticles: Influence of particle size, shape, and chemical surrounding. Phys. Rev. Lett., (2002). , 257404 EOF.

[3] Vartanyan, T., Bosbach, J., Stietz, F., & Träger, F. Theory of spectral hole burning for the study of ultrafast electron dynamics in metal nanoparticles. Appl. Phys. B, (2001). , 73-391.

[4] Bonch-Bruevich, A. M., Vartanyan, T. A., Leonov, N. B., Przhibel'skii, S. G., & Khromov, V. V. Comparative Investigation of the Effect of Heat and Optical Radiation on the Structure of Island Metal Films by Optical Fluctuation Microscopy. Optics and Spectrosc., (2001). , 91(5), 779-785.

[5] Simrick N.J., Kilner J.A., Atkinson A., Thermal stability of silver thin films on zirconia substrates, Thin Solid Films,. (2012). 2012(520), 2855-2867.

[6] Kim H.C., Alford T.L., Allee D.R., Thickness dependence on the thermal stability of silver thin films, Appl.Phys. Lett, (2002). , 4287 EOF-4289 EOF.

[7] Lv J., Lai F., Lin L., Lin Y., Huang Z., Chen R., Thermal stability of Ag films in air prepared by thermal evaporation, Appl.Surf. Sci, (2007). , 2007(253), 7036-7040.

[8] Kaliteevskaya E.N., Krutyakova, V.P., Razumova, T.K., Thermally induced variations in the conformational compositionand spatial orientation of molecular components of a dicarbocyanine dye layer, Opt. and Spectr., (2006). , 300 EOF-306 EOF.

[9] Asnis, L. N., Kaliteevskaya, E. N., Krutyakova, V. P., Razumova, T. K., Tarnovskii, A. N., Tibilov, A. S., & Chizhov, S. A. (2008). , 6985, 6985A-08.

Chapter 4

Lattice Boltzmann Modeling of the Gas Diffusion Layer of the Polymer Electrolyte Fuel Cell with the Aid of Air Permeability Measurements

Hironori Nakajima

Additional information is available at the end of the chapter

1. Introduction

Polymer electrolyte fuel cells (PEFCs, PEMFCs) with high efficiency and low environmental impact recently have attracted considerable interest. However, further improvement in performance and reliability is required to realize practical use of PEFCs as future power generation devices. To improve PEFC performance, an appropriate water balance between the water content and product water is a key technology. Loss of water content in the polymer electrolyte membrane decreases proton conductivity, thereby increasing the internal resistance of the cell. A PEFC basically consists of a membrane electrode assembly (MEA), gas diffusion layers (GDLs) and separators having flow fields with flow channels and ribs. The design parameters for the GDL, such as thickness, pore size distribution, and gas permeability play important roles in characterizing the gas flow and water management during PEFC operation[1]. In this chapter, 2D anisotropic modeling of a monolayer of the GDL substrate is carried out by comparing calculated and measured gas permeability with the lattice Boltzmann method (LBM)[2, 3] and through-plane/in-plane gas permeability measurements, respectively.

2. Lattice Boltzmann method

LBM is a numerical fluid dynamic simulation method that describes macroscopic fluid dynamic phenomena by analyzing the behavior of virtual particles of which fluid is regarded as aggregate. LBM gives simplified kinetic models that incorporate the essential physics of microscopic processes so that the macroscopic averaged properties obey the macroscopic Navier–Stokes equations. Because the conventional Navier–Stokes equation takes long time to calculate and results in poor convergence in porous media, LBM has been developed to take advantage the simplicity of the algorithm and flexibity for complex geometries such as porous media[4]. It is therefore reasonable to apply LBM to fluid flows in the porous structure of the GDLs[5].

2.1. Governing equations

LBM analyzes flow by solving the lattice Boltzmann equation (LBE) that describes particle distribution function, which represents flow velocities of virtual particles. Macroscopic parameters such as the flow velocity and pressure are derived from the summation of the moment of the particle velocity.

In general, LBM uses a single relaxation time approximation by the Bhatnager, Gross, Krook (BGK) model[3, 6]. Equation 1 presents the Boltzmann equation with the BGK approximation

$$\frac{\partial f}{\partial t} + v\nabla f = -\frac{1}{\tau}(f - f^{eq}) \tag{1}$$

where f represents the distribution function depending on space, x, velocity, v and time, t. f^{eq} is the local equilibrium distribution function, and τ is the relaxation time to local equilibrium. The discrete Boltzmann equation is thus

$$\frac{\partial f_\alpha}{\partial t} + v_\alpha \nabla f_\alpha = -\frac{1}{\tau}(f_\alpha - f_\alpha^{eq}) \tag{2}$$

since v-space is discretized by a finite set of particle velocities, v_α and associated distribution function $f_\alpha(x, t)$.

Discretizing with δt and $x + e_\alpha \delta t$, Eq. 2 gives

$$f_\alpha(x + e_\alpha \delta t, t + \delta t) - f_\alpha(x, t) = -\frac{1}{\tau}(f_\alpha(x, t) - f_\alpha^{eq}(x, t)) \tag{3}$$

The collision-streaming process of the LBM is calculated with the following equations.

$$\tilde{f}_\alpha(x, t) = f_\alpha(x, t) - \frac{1}{\tau}(f_\alpha(x, t) - f_\alpha^{eq}(x, t)) \tag{4}$$

$$f_\alpha(x + e_\alpha \delta t, t + \delta t) = \tilde{f}_\alpha(x, t) \tag{5}$$

Here, $\tilde{f}_i$ is the distribution function after the streaming process. The collision process represents the process that the distribution function converges to the equilibrium state, while the streaming process is the process that the virtual particles move to the neighboring sites.

2-dimensional fluid calculation uses the 2D9V model. The virtual particle velocity vector is

$$e = C \begin{bmatrix} 1 & 0 \\ 0 & 1 \\ -1 & 0 \\ 0 & -1 \\ 1 & 1 \\ -1 & 1 \\ -1 & -1 \\ 1 & -1 \\ 0 & 0 \end{bmatrix} \begin{matrix} \alpha = 1 \\ \alpha = 2 \\ \alpha = 3 \\ \alpha = 4 \\ \alpha = 5 \\ \alpha = 6 \\ \alpha = 7 \\ \alpha = 8 \\ \alpha = 9 \end{matrix} \qquad (6)$$

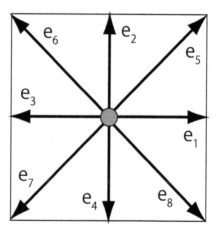

Figure 1. 2D9V velocity model.

where $C = \delta x / \delta t$ is velocity of a virtual particle.

The local distribution function for the 2D9V model is

$$f_\alpha^{eq}(\rho, u) = \omega_\alpha \rho \left(1 + \frac{3}{C^2}(e_\alpha \cdot u) + \frac{9}{2C^4}(e_\alpha \cdot u)^2 - \frac{3}{2C^2}u^2 \right) \qquad (7)$$

where ρ is the density per node, $u = [u_x \ u_y]^T$ is the fluid velocity, ω_α is the weighting function expressed as follows.

Lattice Boltzmann Modeling of the Gas Diffusion Layer of the Polymer Electrolyte Fuel Cell with
the Aid of Air Permeability Measurements

75

$$\omega_{1\sim4} = \frac{1}{9}, \ \omega_{5\sim8} = \frac{1}{36}, \ \omega_9 = \frac{4}{9} \tag{8}$$

The distribution function for each direction is

$$f_1^{eq}(\rho, u) = \frac{1}{9}\rho \left[1 + \frac{3u_x}{C} + \frac{9u_x^2}{2C^2} - \frac{3(u_x^2 + u_y^2)}{2C^2} \right] \tag{9}$$

$$f_2^{eq}(\rho, u) = \frac{1}{9}\rho \left[1 + \frac{3u_y}{C} + \frac{9u_y^2}{2C^2} - \frac{3(u_x^2 + u_y^2)}{2C^2} \right] \tag{10}$$

$$f_3^{eq}(\rho, u) = \frac{1}{9}\rho \left[1 - \frac{3u_x}{C} + \frac{9u_x^2}{2C^2} - \frac{3(u_x^2 + u_y^2)}{2C^2} \right] \tag{11}$$

$$f_4^{eq}(\rho, u) = \frac{1}{9}\rho \left[1 - \frac{3u_y}{C} + \frac{9u_y^2}{2C^2} - \frac{3(u_x^2 + u_y^2)}{2C^2} \right] \tag{12}$$

$$f_5^{eq}(\rho, u) = \frac{1}{36}\rho \left[1 + \frac{3(u_x + u_y)x}{C} + \frac{9u(u_x + u_y)^2}{2C^2} - \frac{3(u_x^2 + u_y^2)}{2C^2} \right] \tag{13}$$

$$f_6^{eq}(\rho, u) = \frac{1}{36}\rho \left[1 + \frac{3(u_x - u_y)}{C} + \frac{9u(u_x - u_y)^2}{2C^2} - \frac{3(u_x^2 + u_y^2)}{2C^2} \right] \tag{14}$$

$$f_7^{eq}(\rho, u) = \frac{1}{36}\rho \left[1 - \frac{3(u_x + u_y)}{C} + \frac{9u(u_x + u_y)^2}{2C^2} - \frac{3(u_x^2 + u_y^2)}{2C^2} \right] \tag{15}$$

$$f_8^{eq}(\rho, u) = \frac{1}{36}\rho \left[1 + \frac{3(u_x - u_y)}{C} + \frac{9u(u_x - u_y)^2}{2C^2} - \frac{3(u_x^2 + u_y^2)}{2C^2} \right] \tag{16}$$

.

$$f_9^{eq}(\rho, \boldsymbol{u}) = \frac{4}{9}\rho\left[1 - \frac{3(u_x^2 + u_y^2)}{2C^2}\right]$$ (17)

The density and macroscopic flow velocity at a node are defined by

$$\rho = \sum f_\alpha$$ (18)

$$\rho\boldsymbol{u} = \sum f_\alpha \boldsymbol{e}_\alpha$$ (19)

respectively.

Equations 6 and 19 give the velocity as follows.

$$u_x = C(f_1 + f_5 + f_8 - f_3 - f_6 - f_7)/\rho$$ (20)

$$u_y = C(f_2 + f_5 + f_6 - f_4 - f_7 - f_8)/\rho$$ (21)

In 2D9V model, the sound velocity C_s and the pressure p are

$$C_s = \frac{C}{\sqrt{3}}$$ (22)

$$p = \rho C_s^2 = \frac{1}{3}\rho C^2$$ (23)

Kinetic viscosity is expressed as

$$\nu = (\tau - \frac{1}{2})C_s^2 \delta t$$ (24)

2.2. Boundary conditions

LBM defines the velocity distribution function at boundary from the velocity and pressure to use the boundary condition. In general, the following bounce back boundary condition has been employed.

2.2.1. Half-way wall bounce-back boundary condition

Half-way wall bounce back boundary condition is no-slip boundary condition at a given solid surface as follows[2, 3].

$$f_1''(x - \delta x, y) = \tilde{f}_3(x, y) \tag{25}$$

$$f_8''(x - \delta x, y + \delta c) = \tilde{f}_6(x, y) \tag{26}$$

$$f_5''(x - \delta x, y - \delta c) = \tilde{f}_7(x, y) \tag{27}$$

where f_α'' represents the distribution function after the streaming step. Since this boundary condition gives higher precision than the conventional bounce back boundary condition[3], it is employed in the present chapter.

2.2.2. Periodic boundary condition

For large area calculation, periodicity of the solution can be assumed. In this case, the periodic boundary condition is employed along the axis direction. The distribution function is

$$f_1''(0, y) = \tilde{f}_1(Nx, y) \tag{28}$$

$$f_5''(0, y) = \tilde{f}_5(Nx, y) \tag{29}$$

$$f_8''(0, y) = \tilde{f}_8(Nx, y) \tag{30}$$

$$f_3''(Nx, y) = \tilde{f}_3(0, y) \tag{31}$$

Before streaming step   After streaming step

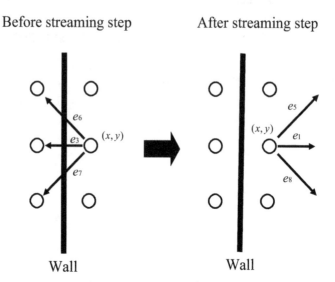

Figure 2. Half-way bounce-back boundary condition

$$f_6''(Nx, y) = \tilde{f}_6(0, y) \tag{32}$$

$$f_7''(Nx, y) = \tilde{f}_7(0, y) \tag{33}$$

where Nx is the maximum value of x.

2.2.3. Pressure difference boundary condition

The pressure difference boundary condition is applied to a case that there is pressure difference between inlet and outlet while the velocity distribution is the same at the inlet and outlet. The distribution functions at the inlet are assumed as follows[7].

$$f_1''(0, y) = \tilde{f}_1(Nx, y) + D \tag{34}$$

$$f_5''(0, y) = \tilde{f}_5(Nx, y) + \frac{1}{4}D \tag{35}$$

Lattice Boltzmann Modeling of the Gas Diffusion Layer of the Polymer Electrolyte Fuel Cell with
the Aid of Air Permeability Measurements

79

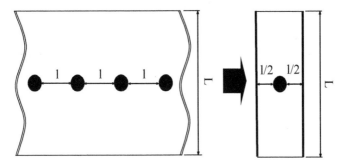

|: Periodic boundary condition

Figure 3. Periodic boundary condition.

$$f_8''(0, y) = \tilde{f}_8(Nx, y) + \frac{1}{4}D \tag{36}$$

with

$$D = \frac{\Delta p}{C^2} - \frac{1}{3}\left[\tilde{f}_2(1, y) - \tilde{f}_2(Nx, y) + \tilde{f}_4(1, y) - \tilde{f}_4(Nx, y) + \tilde{f}_9(1, y) - \tilde{f}_9(Nx, y)\right] \tag{37}$$

while the distribution functions at the outlet are

$$f_3''(Nx, y) = \tilde{f}_3(0, y) - D \tag{38}$$

$$f_6''(Nx, y) = \tilde{f}_6(0, y) - \frac{1}{4}D \tag{39}$$

$$f_7''(Nx, y) = \tilde{f}_7(0, y) - \frac{1}{4}D \tag{40}$$

2.2.4. Pressure and velocity boundary conditions

Pressure and velocity boundary conditions proposed by Zou and He[8] are also used for the calculation. The case for an inlet at $x = 0$ is considered for instance here. At the boundary, pressure, that is, ρ_{in}, and $u_y = 0$ are applied. Because f_2'', f_3'', f_4'', f_6'', f_7'', f_9'' after streaming step are known, u_x, f_1'', f_5'', f_8'' are derived as follows.

Equation 18 gives

$$f_1'' + f_5'' + f_8'' = \rho_{in} - (f_2'' + f_3'' + f_4'' + f_6'' + f_7'' + f_9'') \tag{41}$$

while Eqs. 20 and 21 lead to

$$C(f_1'' + f_5'' + f_8'') = \rho_{in}u_x + C(f_3'' + f_6'' + f_7'') \tag{42}$$

$$f_5'' - f_8'' = -f_2'' + f_4'' - f_6'' + f_7'' \tag{43}$$

Hence

$$u_x = C\left[1 - \frac{f_2'' + f_4'' + f_9'' + 2(f_3'' + f_6'' + f_7'')}{\rho_{in}}\right] \tag{44}$$

Then the deviation from the equilibrium shall be equal for the distribution function of $i = 1,3$ as follows to determine the remaining distribution functions.

$$f_1'' - f_1^{eq} = f_3'' - f_3^{eq} \tag{45}$$

Thus Eqs. 9 and 11 yield

$$f_1'' = f_3'' + \frac{2\rho u_x}{3C} \tag{46}$$

Lattice Boltzmann Modeling of the Gas Diffusion Layer of the Polymer Electrolyte Fuel Cell with
the Aid of Air Permeability Measurements

81

Eqs. 42, 43, and 46 give

$$f_5'' = f_7'' - \frac{f_2'' - f_4''}{2} + \frac{\rho u_x}{6C} \tag{47}$$

$$f_8'' = f_6'' - \frac{f_2'' - f_4''}{2} + \frac{\rho u_x}{6C} \tag{48}$$

Thereby u_x, f_1, f_5, f_8 are determined. On the other hand, the distribution function at the corner should be dealt with in other way. The case for the bottom of the inlet at $x = 0, y = 0$ is described here for instance. After the streaming step, f_3'', f_4'', f_7'', ρ_{in} is obtained. The no-slip boundary condition gives $u_x = 0$, $u_y = 0$. Thus f_1, f_2, f_5, f_6, f_8 can be determined as follows. Eq. 46 provides

$$f_1'' = f_3'' \tag{49}$$

In a similar manner,

$$f_2'' = f_4'' \tag{50}$$

Equations 18 and 47 yield

$$f_5'' = f_7'' \tag{51}$$

$$f_6'' = f_8'' = \frac{1}{2}\left[\rho_{in} - (f_1'' + f_2'' + f_3'' + f_4'' + f_5'' + f_7'' + f_9'')\right] + \tag{52}$$

From Eq. 44,

$$\rho = \frac{C}{C - u_x}\left[f_2'' + f_4'' + f_9'' + 2(f_3'' + f_6'' + f_7'')\right] \tag{53}$$

2.3. LBM binary mixtures with different molecular weights

In this section, LBM binary mixtures with different molecular weights (LBM-BMD) model proposed by Luo and Girimaji[9] and extended by McCracken and Abraham[10] is described. LMB-BMD model consists of LBM and the effect of diffusion. This model deals with two components A and B having different molecular weights. The model can analyze advective flow in addition to diffusion. i and j represent the functions, variables, and constants for the species A and B, respectively.

The equilibrium distribution function is

$$f_\alpha^i(x + e_\alpha^i \delta t, t + \delta t) - f_{\alpha i}^i(x, t) = \Omega_\alpha^{ii} + \Omega_\alpha^{ij} \tag{54}$$

where Ω_α^{ii} and Ω_α^{ij} are the following self-collision and cross-collision terms for A-A and A-B, respectively. Velocity vector is defined as the similar manner as LBM in the previous section.

$$\Omega_\alpha^{ii} = -\frac{1}{\tau^i}(f_\alpha^i(x, t) - f_\alpha^{i(0)}(x, t)) \tag{55}$$

$$\Omega_\alpha^{ij} = -\frac{1}{\tau_D^{ij}} \left(\frac{\rho_j}{\rho}\right) \frac{f_\alpha^{i(0)}}{(C_s^i)^2}(e_\alpha^i - u)(u^i - u^j) \tag{56}$$

where τ^i and τ_D^{ij} are the relaxation times for the kinetic viscosity, ν^i, and diffusion coefficient, D^{ij}. The local equilibrium distribution function, $f_\alpha^{i(0)}$ is

$$f_\alpha^{i(0)} = \left[1 + \frac{3}{C^{i2}}(e_\alpha^i - u)(u^i - u^j)\right] f_\alpha^{i,eq} \tag{57}$$

In a similar manner as the single component LBM in the previous section,

$$f_\alpha^{eq}(\rho, u) = \omega_\alpha \rho^i (1 + \frac{3}{C^{i2}}(e_i \cdot u) + \frac{9}{2C^{i4}}(e_i \cdot u)^2 - \frac{3}{2C^{i2}}u^2) \tag{58}$$

$$f_1^{i(0)} = \left\{1 + \frac{3}{C^{i2}}\left[(C^i - u_x)u_{diff,x}^i - u_y u_{diff,y}^i\right]\right\} f_1^{i,eq} \tag{59}$$

Lattice Boltzmann Modeling of the Gas Diffusion Layer of the Polymer Electrolyte Fuel Cell with
the Aid of Air Permeability Measurements

83

$$f_2^{i(0)} = \left\{1 + \frac{3}{C^{i2}}\left[(C^i - u_y)u_{\text{diff},y}^i - u_x u_{\text{diff},x}^i\right]\right\} f_2^{i,\text{eq}} \tag{60}$$

$$f_3^{i(0)} = \left\{1 - \frac{3}{C^{i2}}\left[(C^i - u_x)u_{\text{diff},x}^i + u_y u_{\text{diff},y}^i\right]\right\} f_3^{i,\text{eq}} \tag{61}$$

$$f_4^{i(0)} = \left\{1 - \frac{3}{C^{i2}}\left[(C^i + u_y)u_{\text{diff},y}^i + u_x u_{\text{diff},x}^i\right]\right\} f_4^{i,\text{eq}} \tag{62}$$

$$f_5^{i(0)} = \left\{1 + \frac{3}{C^{i2}}\left[(C^i - u_x)u_{\text{diff},x}^i + (C^i - u_y)u_{\text{diff},y}^i\right]\right\} f_5^{i,\text{eq}} \tag{63}$$

$$f_6^{i(0)} = \left\{1 + \frac{3}{C^{i2}}\left[(C^i - u_x)u_{\text{diff},x}^i + (C^i - u_y)u_{\text{diff},y}^i\right]\right\} f_6^{i,\text{eq}} \tag{64}$$

$$f_7^{i(0)} = \left\{1 + \frac{3}{C^{i2}}\left[(C^i - u_x)u_{\text{diff},x}^i - (C^i - u_y)u_{\text{diff},y}^i\right]\right\} f_7^{i,\text{eq}} \tag{65}$$

$$f_8^{i(0)} = \left\{1 + \frac{3}{C^{i2}}\left[(C^i - u_x)u_{\text{diff},x}^i - (C^i + u_y)u_{\text{diff},y}^i\right]\right\} f_8^{i,\text{eq}} \tag{66}$$

$$f_9^{i(0)} = \left\{1 - \frac{3}{C^{i2}}\left[u_x u_{\text{diff},x}^i + u_y u_{\text{diff},y}^i\right]\right\} f_9^{i,\text{eq}} \tag{67}$$

where $u_{\text{diff},x}^i$ and $u_{\text{diff},y}^i$ are x and y direction of $\boldsymbol{u}_{\text{diff}}^i = \boldsymbol{u}^i - \boldsymbol{u}$

Collision terms are:

$$\Omega_1^{ij} = -\frac{3}{\tau_D^{ij}} \left(\frac{\rho^j}{\rho} \right) \frac{f_1^{i(0)}}{C^{i2}} \left[(C^i - u_x)u_x^{i-j} - u_y u_y^{i-j} \right] \tag{68}$$

$$\Omega_2^{ij} = -\frac{3}{\tau_D^{ij}} \left(\frac{\rho^j}{\rho} \right) \frac{f_2^{i(0)}}{C^{i2}} \left[(C^i - u_y)u_y^{i-j} - u_x u_x^{i-j} \right] \tag{69}$$

$$\Omega_3^{ij} = \frac{3}{\tau_D^{ij}} \left(\frac{\rho^j}{\rho} \right) \frac{f_3^{i(0)}}{C^{i2}} \left[(C^i + u_x)u_x^{i-j} + u_y u_y^{i-j} \right] \tag{70}$$

$$\Omega_4^{ij} = \frac{3}{\tau_D^{ij}} \left(\frac{\rho^j}{\rho} \right) \frac{f_3^{i(0)}}{C^{i2}} \left[(C^i + u_y)u_y^{i-j} + u_x u_x^{i-j} \right] \tag{71}$$

$$\Omega_5^{ij} = -\frac{3}{\tau_D^{ij}} \left(\frac{\rho^j}{\rho} \right) \frac{f_3^{i(0)}}{C^{i2}} \left[(C^i - u_x)u_x^{i-j} + (C^i - u_y)u_y^{i-j} \right] \tag{72}$$

$$\Omega_6^{ij} = -\frac{3}{\tau_D^{ij}} \left(\frac{\rho^j}{\rho} \right) \frac{f_3^{i(0)}}{C^{i2}} \left[(C^i - u_x)u_x^{i-j} + (C^i - u_y)u_y^{i-j} \right] \tag{73}$$

$$\Omega_7^{ij} = -\frac{3}{\tau_D^{ij}} \left(\frac{\rho^j}{\rho} \right) \frac{f_3^{i(0)}}{C^{i2}} \left[(C^i - u_x)u_x^{i-j} + (C^i - u_y)u_y^{i-j} \right] \tag{74}$$

$$\Omega_8^{ij} = -\frac{3}{\tau_D^{ij}} \left(\frac{\rho^j}{\rho} \right) \frac{f_3^{i(0)}}{C^{i2}} \left[(C^i - u_x)u_x^{i-j} + (C^i - u_y)u_y^{i-j} \right] \tag{75}$$

where

Lattice Boltzmann Modeling of the Gas Diffusion Layer of the Polymer Electrolyte Fuel Cell with
the Aid of Air Permeability Measurements

85

$$u^{i-j} = \begin{bmatrix} u_x^{i-j} \\ u_y^{i-j} \end{bmatrix} = \begin{bmatrix} u_x^i - u_x^j \\ u_y^i - u_y^j \end{bmatrix} \tag{76}$$

The density and flow velocity are derived by

$$\rho^i = \sum f_\alpha^i \tag{77}$$

$$\rho^i u^i = \sum f_\alpha^i e_\alpha^i \tag{78}$$

The total density and mass averaged velocity are

$$\rho = \rho^i + \rho^j \tag{79}$$

$$\rho u = \rho^i u^i + \rho^j u^j \tag{80}$$

Since the partial pressure is

$$p_i = \rho_i C_s^{i2} \tag{81}$$

total pressure is

$$p = \rho^i C_s^{i2} + \rho^j C_s^{j2} = \frac{1}{3}(\rho^i C_s^{i2} + \rho^j C_s^{j2}) \tag{82}$$

The relation between the sound velocities, C_s^i and C_s^j is

$$C_s^j = \sqrt{\frac{m^i}{m^j}} C_s^i \tag{83}$$

where $m^i > m^j$ are the molecular weights of A and B, respectively. The kinetic viscosity and diffusion coefficient have the following relations.

$$\nu^i = \frac{1}{3}(\tau^i - \frac{1}{2})C_s^{i2}\delta t \tag{84}$$

$$D^{ij} = \frac{\rho p}{n^2 m^i m^j}\left(\tau_D^i - \frac{1}{2}\right) \tag{85}$$

where

$$n^i = \frac{\rho^i}{m^i} , \; n^j = \frac{\rho^j}{m^j} \tag{86}$$

$$n = n^i + n^j \tag{87}$$

2.4. Streaming step of species with different velocities

In LBM-BMD model, species A snd B have different velocities. During δt , the species A travel δx, while the species B travel $\sqrt{m^i/m^j}\delta x$. Since $\delta x = e_\alpha^i \delta t$, the species B have different streaming distance. So, the distribution function of B on the nodes should be determined from the interpolation of the distribution functions of surrounding particles. Although McCracken and Abraham proposed a second-order Lagrangian interpolation[10], and Joshi et. al proposed bi-linear interpolation[11], these interpolation methods seem not appropriate for porous structure despite their higher accuracy. Thus linear interpolation of fewer nodes is employed here. The case for $\alpha = 1$ is depicted in Fig. 4. After the streaming step, the distribution function at (x, y) is determined by the interpolation of the distribution functions at $(x - \delta x, y)$ and (x, y) before the streaming step.

Lattice Boltzmann Modeling of the Gas Diffusion Layer of the Polymer Electrolyte Fuel Cell with
the Aid of Air Permeability Measurements

87

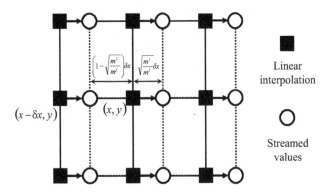

Figure 4. Streaming and interpolation for species B

$$f_1^{j''}(x, y) = \tilde{f}_1^j(x - \delta x, y)$$

$$+ \frac{x - \left[(x - \delta x) + \sqrt{\frac{m^i}{m^j}}\delta x\right]}{\left(\sqrt{\frac{m^i}{m^j}}\delta x\right) - \left[(x - \delta x) + \sqrt{\frac{m^i}{m^j}}\delta x\right]} \left[\tilde{f}_1^j(x, y) - \tilde{f}_1^j(x - \delta x, y)\right] \tag{88}$$

$$= \sqrt{\frac{m^i}{m^j}}\tilde{f}_1^j(x - \delta x, y) + \left(1 - \sqrt{\frac{m^i}{m^j}}\right)\tilde{f}_1^j(x, y) \tag{89}$$

when $(x - \delta x, y)$ is an obstacle node, $\tilde{f}_1^j(x - \delta x, y) = \tilde{f}_3^j(x, y)$ can be applied.

$$f_1^{j''}(x, y) = \sqrt{\frac{m^i}{m^j}}\tilde{f}_3^j(x, y) + \left(1 - \sqrt{\frac{m^i}{m^j}}\right)\tilde{f}_1^j(x, y) \tag{90}$$

3. GDL models

In the present research, 3D structure of the GDL is projected to 2D structure. So, the following models are created.

Model 1 Cross-section of the carbon fiber is simulated as a circle so that averaged number of the fiber in unit area is the same as that of the actual GDL.

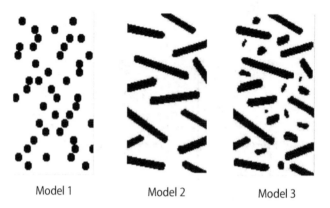

Model 1　　　　　Model 2　　　　　Model 3

Figure 5. 2D anisotropic GDL models for the LBM.

Model 2 Fiber is simulated so that porosity is the same as the actual GDL.

Model 3 Fiber and its cross-section are simulated by so that porosity is the same as the actual GDL.

Figure 5 illustrates the GDL models. Through-plane and in-plane anisotropic Darcy coefficients are obtained by LBM employed to these GDL models. These Darcy coefficients are compared with those from the following permeability measurements so that an anisotropic GDL model which agrees the most with the measurements can be found. The GDL model found is then used for the LBM-BMD flow analysis in the GDL with the flow channels and ribs having actual PEFC flow field geomeory under actual operation condition.

The calculation in the present chapter was carried out with a personal computer having Intel Core2 Quad CPU Q6600 2.4GHz and 4GB memory on ASUS P5K motherboard. MatLab (MathWorks, Inc.) was used for the LBM and LBM-BMD calculations.

4. Experimental

Figure 6 shows a schematic diagram of GDL permeability measurement apparatus. GDL, which was a commercial carbon paper (SIGRACET GDL 24AA, SGL Carbon Inc.) with a thickness of 190 μm, was placed between two cylindrical plates. A soft O-ring was used for gas sealing between the plates. The force required to deform the O-ring was negligible compared with the compression force acting on the GDL. The compression force was controlled using a clamp screw and was measured with a load cell. For air permeability tests, the compression pressure was set at 1 MPa, as measured in a typical PEFC.

Fig. 7 presents geometries of the GDL used for the through-plane and in-plane permeability tests[12]. Volumetric air flow rates in through-plane direction, Qth, and in-plane direction, Qin, in the following equations were measured using a mass flow meter (KOFLOC). Pressure drop by the apparatus was compensated beforehand.

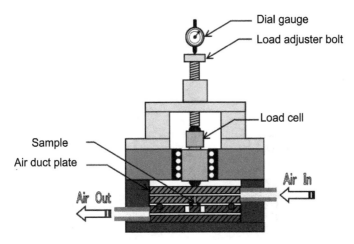

Figure 6. Apparatus for the permeability measurement.

$$Q_{th} = \frac{k}{\mu} \frac{p_i - p_o}{\delta} A = \frac{k}{\mu} \frac{p_i - p_o}{\delta} 2\pi r^2 \qquad (91)$$

$$Q_{in} = \frac{k}{\mu} \frac{p_i - p_o}{r_o - r_i} \frac{2\pi t (r_o - r_i)}{\ln(r_o / r_i)} \qquad (92)$$

where k, μ, and r are the Darcy coefficient of the GDL, viscosity of air[13], and radius of the GDL, respectively. P_i, P_O, δ, and A are inlet and outlet air pressures, thickness of the GDL, and cross-sectional area of air flow, respectively. r_i and r_O are inner and outer radii of the GDL for the in-plane permeability measurement. The Darcy coefficients in through-plane and in-plane directions are thereby obtained from relations between the flow rates and the pressure difference.

5. Results and discussion

Figures 8 and 9 show the through-plane and in-plane Darcy coefficients obtained from the LBM calculation and permeability measurements. Darcy coefficients of the GDL model 3 agrees well with the experimental results in the cases of in-plane flow and through-plane flow below flow velocity of 1 ms^{-1}. Since the through-plane flow velocity is below 1 ms^{-1} in a cell in general, the model 3 shall be used for the LBM-BMD calculation for GDL under flow channels and a rib in an actual cell below.

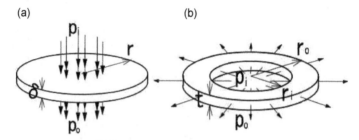

Figure 7. GDL geometries for the permeability measurements. (a)Through-plane (b)In-plane

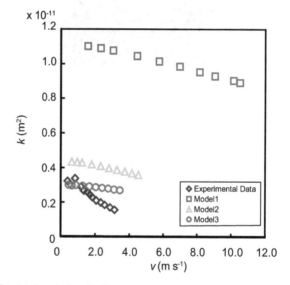

Figure 8. Darcy coefficients in through-plane direction.

Figure 10 illustrates a parallel-serpentine flow field in the cathode of a PEFC. Flow analyses for the GDL between the flow channels indicated in the blue and red circles. The pressure difference between the channels in the former part is rather smaller than the lattar. This comparison represents that between the parallel and serpentine flow fields.

The GDL is modeled with the flow channel and rib as presented in Fig. 11.

Cell temperature is 75°C and water vapor pressure shall be the saturation vapor pressure at 75 °C. Air utilization is 30%. Oxygen and nitrogen partial pressure is assumed to linearly change from the inlet to the outlet, yielding 1.60 and -0.81 kPa, respectively, between the flow channels in the red part at 1.0 A cm^{-2}. It is assumed that current distribution is uniform and there is no liquid water in the GDL to simplify the calculation. Thickness of the GDL, rib width, and channel width are 190 μm, 0.6 mm, and 0.5 mm, respectively. Viscosities of

Lattice Boltzmann Modeling of the Gas Diffusion Layer of the Polymer Electrolyte Fuel Cell with
the Aid of Air Permeability Measurements

91

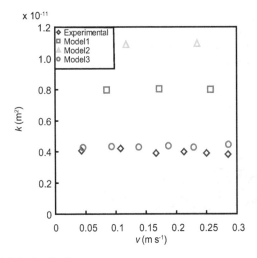

Figure 9. Darcy coefficients in in-plane direction.

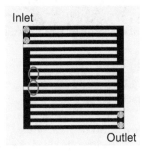

Figure 10. Serpentine flow field.

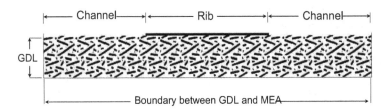

Figure 11. GDL model with the channels and rib of the PEFC.

nitrogen and oxygen of 20.00×10^{-6} and $23.26 \times \times 10^{-6}$ Pa·s, respectively[13], and binary diffusion coefficient for nitrogen-oxygen mixture of $2.59 \times 10^{-5} m^2 s^{-1}$[14] are used for the LBM-BMD calculation. Oxygen flow velocity distribution at 1.0 A cm^{-2} is presented in Fig. 12. Oxygen amount consumed by the electrochemical reaction is calculated with the Faraday's law.

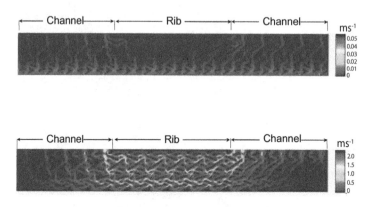

Figure 12. Oxygen flow velocity distributions in the GDLs (a) without pressure difference and (b) with pressure difference between the flow channels at 75 °C, 1.0 A cm^{-2}.

Figures 12(a) and (b) depict oxygen flow velocities for the GDLs without and with pressure differences between the flow channels, respectively. Oxygen is transported to the surface of the MEA mainly by diffusion in the case without the pressure differences since there is small forced convection thorough the GDL. On the other hand, oxygen flow velocity is rather larger under the rib in the case with pressure difference that leads to forced convection in the GDL. The forced convection also enhances the discharge of liquid and vapor product water in actual cells. The forced convection in the GDL by the pressure difference between flow channels plays a significant role on the exhaust of the product water and oxygen transport for the interdigitated flow field[15, 16].

6. Conclusion

In this chapter, an anisotropic 2D GDL model is proposed by comparisons of the through-plane and in-plane permeabilities between those obtained by LBM calculation and permeability measurements. The modeled carbon fiber structure agrees well with the actual GDL in terms of the permeability. Moreover, the difference of oxygen flow in GDLs with parallel and serpentine flow channels is visualized with the oxygen-nitrogen two components LBM-BMD calculations using the above anisotropic GDL model. This procedure can be used to optimize the GDL porous structure, flow field patterns, and operation conditions of PEFCs. Liquid-gas two phase modeling[17, 18], modeling, modeling of microporous layers[19–21], and expansion to 3D modelings are future studies.

Acknowledgments

The author is grateful to a graduate student, Tomokazu KOBAYASHI (presently TOYOTA Motor Corp.) for considerable assistance with the modeling and calculation. The author also thank Associate Professor Kitahara and a graduate student, Teppei YASUKAWA (presently DENSO Corp.) for help with the permeability measurements. The author is indebted to Professor Konomi for valuable discussions.

Author details

Hironori Nakajima

Department of Mechanical Engineering at Kyushu University, Japan

References

[1] H. Nakajima, T. Konomi, and T. Kitahara. Direct water balance analysis on a polymer electrolyte fuel cell (PEFC): Effects of hydrophobic treatment and micro porous layer addition to the gas diffusion layer of a PEFC on its performance during a simulated start-up operation. *Journal of Power Sources*, 171:457–463, 2007.

[2] Dieter A. Wolf-Gladrow. *Lattice-Gas Cellular Automata and Lattice Boltzmann Models: An introduction*, volume 1725 of *Lecture Notes in Mathematics*. Springer, 2000.

[3] Sauro Succi. *The Lattice Boltzmann Equation for Fluid Dynamics and Beyond*. Oxford University Press, New York, 2001.

[4] M. Yoshino and T. Inamuro. Lattice boltzmann simulations for flow and heat/mass transfer problems in a three-dimensional porous structure. *International Journal for Numerical Methods in Fluids*, 43(2):183–198, 2003.

[5] J. Park and X. Li. Multi-phase micro-scale flow simulation in the electrodes of a PEM fuel cell by lattice Boltzmann method. *Journal of Power Sources*, 178(1):248 – 257, 2008.

[6] P. L. Bhatnagar, E. P. Gross, and M. Krook. A model for collision processes in gases. i. small amplitude processes in charged and neutral one-component systems. *Phys. Rev.*, 94:511–525, May 1954.

[7] Takaji Inamuro, Koji Maeba, and Fumimaru Ogino. Flow between parallel walls containing the lines of neutrally buoyant circular cylinders. *International Journal of Multiphase Flow*, 26(12):1981 – 2004, 2000.

[8] Qisu Zou and Xiaoyi He. On pressure and velocity boundary conditions for the lattice Boltzmann BGK model. *Physics of Fluids*, 9(6):1591–1598, 1997.

[9] Li-Shi Luo and Sharath S. Girimaji. Lattice Boltzmann model for binary mixtures. *Physical Review E*, 66:035301, Sep 2002.

[10] Michael E. McCracken and John Abraham. Lattice Boltzmann methods for binary mixtures with different molecular weights. *Physical Review E*, 71:046704, Apr 2005.

[11] Abhijit S Joshi, Aldo A Peracchio, Kyle N Grew, and Wilson K S Chiu. Lattice boltzmann method for continuum, multi-component mass diffusion in complex 2D geometries. *Journal of Physics D: Applied Physics*, 40(9):2961, 2007.

[12] Tatsumi Kitahara, Toshiaki Konomi, and Hironori Nakajima. Microporous layer coated gas diffusion layers for enhanced performance of polymer electrolyte fuel cells. *Journal of Power Sources*, 195(8):2202 – 2211, 2010.

[13] *JSME Data Book: Thermophysical Properties of Fluids*. The Japan Society of Mechanical Engineers, Tokyo, 1983.

[14] S. Oe. *Estimation Methods of Physical Property Constants for Designers (in Japanese)*. Nikkan Kogyo Shimbun, Tokyo, 1985.

[15] T. V. Nguyen. A gas distributor design for proton-exchange-membrane fuel cells. *Journal of the Electrochemical Society*, 143:L103–L105, 1996.

[16] D.L. Wood III, J.S. Yi, and T.V. Nguyen. Effect of direct liquid water injection and interdigitated flow field on the performance of proton exchange membrane fuel cells. *Electrochimica Acta*, 43(24):3795–3809, 1998.

[17] U. Pasaogullari and C.-Y. Wang. Two-phase transport and the role of micro-porous layer in polymer electrolyte fuel cells. *Electrochimica Acta*, 49(25):4359–4369, 2004.

[18] U. Pasaogullari and C.Y. Wang. Liquid water transport in gas diffusion layer of polymer electrolyte fuel cells. *Journal of the Electrochemical Society*, 151(3):A399–A406, 2004.

[19] Z. Qi and A. Kaufman. Improvement of water management by a microporous sublayer for PEM fuel cells. *Journal of Power Sources*, 109(1):38–46, 2002.

[20] A.Z. Weber and J. Newman. Effects of microporous layers in polymer electrolyte fuel cells. *Journal of the Electrochemical Society*, 152(4):A677–A688, 2005.

[21] Tatsumi Kitahara, Hironori Nakajima, and Kyohei Mori. Hydrophilic and hydrophobic double microporous layer coated gas diffusion layer for enhancing performance of polymer electrolyte fuel cells under no-humidification at the cathode. *Journal of Power Sources*, 199:29 – 36, 2012.

Chapter 5

Airlift Bioreactors: Hydrodynamics and Rheology Application to Secondary Metabolites Production

Ana María Mendoza Martínez and
Eleazar Máximo Escamilla Silva

Additional information is available at the end of the chapter

1. Introduction

The bubble column and airlift bioreactors are pneumatically agitated and often employed in bioprocesses where gas-liquid contact is important. The role of the gas is to provide contact with the liquid for mass transfer processes such as absorption or desorption and to provide energy through gas expansion or bubble buoyancy for liquid mixing. In these two pneumatically agitated reactors, gas is sparger usually through the bottom and the buoyancy of the ascending gas bubbles causes mixing. The main difference between these two pneumatically agitated reactors is in their fluid flow characteristics. The flow in the airlift is ordered and in a cyclic pattern like in a loop beginning from top through to bottom. The airlift differs from the bubble column by the introduction of inner draft tubes which improves circulation, whereas the bubble column is a simple tower. In the airlift, liquid recirculation occurs due to the four distinct sections; the riser, downcomer, gas separator and bottom or base. The bubble column is a simple vessel without any sectioning making the flow rather a complex one.

Some attractive features of the airlift are the low power consumption, simplicity in construction with no moving parts, high mass and heat transfer rates and uniform distribution of shear [1, 2].

The advantage of its low power consumption is of particular importance in effluent (e.g. wastewater) treatment where the product value is comparatively low. Therefore, operational cost (efficient use of energy) is greatly considered since corresponding applications are usually on a large scale. Homogenous shear is particularly important for biological processes that are shearing sensitive. In the conventional stirred tank, shear is greatest at the stirrer and decreases away from it to the walls of the vessel. This creates a gradient of shearing which can have adverse effect on the morphology or sometimes can damage cells (e.g. animal and plant cells). The simple construction of the airlift without shafts makes it not only aesthetically pleasing to look at but also eliminates contamination associated with the conventional stirred tank which is a major drawback in the production of microorganism. A sterile environment is crucial for growing organisms especially in the bioprocesses since contamination reduces product quality, generates wastes, also more time and money are spent to restore the whole process.

In addition to the previously described, the development of new biotech products: drugs, vaccines, tissue culture, agrochemicals and specialty chemicals, biofuels and others have had a major breakthrough during the last decade. But engineering to scale these developments to the production phase is lagging far behind in advances in the efficient development of bioprocesses. Normally these processes are complicated because they are conducted in complex systems, three or four phases, microorganisms that are susceptible to large shear produced by stirring; require high airflow rates, and changes in rheology and morphology cultures through the time. The soul of a bioprocess is still the bioreactor, since it determines the success of a good separation and therefore the cost of the product. In this chapter we describe one of the most promising bioreactors for their processing qualities, good mixing, low shear, easy to operate with immobilized microorganisms, low consumption of energy, we are talking about the airlift bioreactors. In this context the parties will start from basic engineering bioreactors: mass balances, mass transfer, and modelling and to cover the hydrodynamics and rheology of the process. Finally we will present two cases of application on the production of Bikaverin (a new antibiotic), and L-lysine.

2. Hydrodynamic characteristic of airlift reactors

Fluid mixing is influenced by the mixing time and gas holdup which defines the fluid circulation and mass transfer properties. The fluid recirculation causes the difference in hydrostatic pressure and density due to partial or total gas disengagement at the gas separator (top clearance, tt). Studies have been documented during the last two decades with various correlations applicable for hydrodynamic parameters [3-5]. This implies that for a successful design, fundamental understanding of mixing parameters is important for industrial scale-up.

It is difficult to generalize the performance of the bioreactor according to the process for which the airlift will be employed. For example, in aerobic fermentation, oxygen is important for mass transfer and therefore, it is imperative to consider a design where there will be less disengagement of gas resulting in higher gas holdup for a higher mass transfer rate. In this case, the liquid circulation velocity is low because less gas is disengaged at the top resulting in a lower differential density. Furthermore, other processes require good mixing other than a high mass transfer rate. However, provision can be made by increasing the gas disengagement at the top to improve the liquid recirculation as in the case for anaerobic fermentation. Therefore, it is safe to conclude as have been confirmed [3, 6, 7] that, the geometry parameters such as the top clearance (tt), ratio of cross sectional area of the downcomer to the riser (Ad/Ar), bottom clearance (tb), the cross sectional areas of riser (Ar) and that of the downcomer

(*Ad*), draft tube internal diameter (*Dd*), and height of the column (*H*) and superficial gas velocity *(Ug)* have an influence on fluid hydrodynamics.

There is extensive information on the measurement of fluid hydrodynamics published with a handful of equations. However, most of them Gumery et al.: Characteristics of Macro-Mixing in Airlift Column Reactors 7 Published by The Berkeley Electronic Press, 2009 cannot be correlated due to the different medium (Newtonian versus non-Newtonian) used and various assumptions made. The different measuring techniques often used cannot discriminate diffusion and convection for mixing whiles others disturb process flow.

On the other hands the interconnections between the design variables, the operating variables, and the observable hydrodynamic variables in an airlift bioreactor are presented diagrammatically in Figure 4 as has been reported by [8]. The design variables are the reactor height, the riser-to-downcomer area ratio, the geometrical design of the gas separator, and the bottom clearance (C_b, the distance between the bottom of the reactor and the lower end of the draft tube, which is proportional to the free area for flow in the bottom and represents the resistance to flow in this part of the reactor). The main operating variables are primarily the gas input rate and, to a lesser extent, the top clearance (C_t, the distance between the upper part of the draft tube and the surface of the non-aerated liquid). These two independent variables set the conditions that determine the liquid velocity in the airlift bioreactor via the mutual influences of pressure drops and holdups, as shown in Figure 1 [9]. Viscosity is not shown in Figure 1 as an independent variable because in the case of gas–liquid mixtures, it is a function of the gas holdup (and of liquid velocity in the case of non-Newtonian liquids), and because in a real process, it will change with time due the changes in the composition of the liquid.

3. Flow configuration

3.1. Riser

In the riser, the gas and liquid flow upward, and the gas velocity is usually larger than that of the liquid. The only exception is homogeneous flow, in which case both phases flow at the same velocity. This can happen only with very small bubbles, in which case the free-rising velocity of the bubbles is negligible with respect to the liquid velocity. Although about a dozen different gas–liquid flow configurations have been developed [10], only two of them are of interest in airlift bioreactors [11, 12]:

1. Homogeneous bubbly flow regime in which the bubbles are relatively small and uniform in diameter and turbulence is low.

2. Churn-turbulent regime, in which a wide range of bubble sizes coexist within a very turbulent liquid.

The churn-turbulent regime can be produced from homogeneous bubbly flow by increasing the gas flow rate. Another way of obtaining a churn-turbulent flow zone is by starting from slug flow and increasing the liquid turbulence, by increasing either the flow rate or the diameter of the reactor, as can be seen in Figure 5 [12]. The slug-flow configuration is important only as a situation to be avoided at all costs, because large bubbles bridging the entire tower cross-section offer very poor capacity for mass transfer.

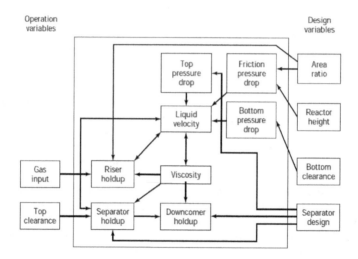

Figure 1. Interaction between geometric and fluid dynamic variables in an airlift bioreactor [9].

3.2. Downcomer

In the downcomer, the liquid flows downward and may carry bubbles down with it. For bubbles to be entrapped and flow downward, the liquid velocity must be greater than the free-rise velocity of the bubbles. At very low gas flow input, the liquid superficial velocity is low, practically all the bubbles disengage, and clear liquid circulates in the downcomer. As the gas input is increased, the liquid velocity becomes sufficiently high to entrap the smallest bubbles. Upon a further increase in liquid velocity larger bubbles are also entrapped. Under these conditions the presence of bubbles reduces the cross-section available for liquid flow, and the liquid velocity increases in this section.

Bubbles are thus entrapped and carried downward, until the number of bubbles in the cross-section decreases, the liquid velocity diminishes, and the drag forces are not sufficient to overcome the buoyancy. This feedback loop in the downcomer causes stratification of the bubbles, which is evident as a front of static bubbles, from which smaller bubbles occasionally escape downward and larger bubbles, produced by coalescence, escape upward. The bubble front descends, as the gas input to the system is increased, until the bubbles eventually reach the bottom and recirculate to the riser. When this point is reached, the bubble distribution in the downcomer becomes much more uniform. This is the most desirable flow configuration in the downcomer, unless a single pass of gas is required. The correct choice of cross-sectional area ratio of the riser to the downcomer will determine the type of flow.

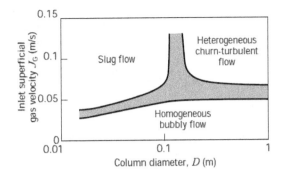

Figure 2. Map of flow configurations for gas–liquid concurrent flow in a vertical tube [12].

3.3. Gas separator

The gas separator is often overlooked in descriptions of experimental airlift bioreactor devices, although it has considerable influence on the fluid dynamics of the reactors. The geometric design of the gas separator will determine the extent of disengagement of the bubbles entering from the riser. In the case of complete disengagement, clear liquid will be the only phase entering the downcomer. In the general case, a certain fraction of the gas will be entrapped and recirculated. Fresh gas may also be entrapped from the headspace if the fluid is very turbulent near the interface. The extent of this entrapment influences strongly gas holdup and liquid velocity in the whole bioreactor.

It is quite common to enlarge the separator section to reduce the liquid velocity and to facilitate better disengagement of spent bubbles. Experiments have been reported in which the liquid level in the gas separator was high enough to be represented as two mixed vessels in series [13]. This point will be analysed further in the section devoted to mixing.

4. Gas holdup

Gas holdup is the volumetric fraction of the gas in the total volume of a gas–liquid–solid dispersion:

$$\varphi_1 = \frac{V_G}{V_L + V_G + V_S} \tag{1}$$

where the sub-indexes L, G, and S indicate liquid, gas, and solid, and i indicates the region in which the holdup is considered, that is, gas separator (s) the riser (r), the downcomer (d), or the total reactor (T).

The importance of the holdup is twofold: (a) the value of the holdup gives an indication of the potential for mass transfer, since for a given system a larger gas holdup indicates a larger gas–

liquid interfacial area; and (b) the difference in holdup between the riser and the downcomer generates the driving force for liquid circulation. It should be stressed, however, that when referring to gas holdup as the driving force for liquid circulation, only the total volume of the gas is relevant. This is not the case for mass transfer phenomena, in this case, the interfacial area is of paramount importance, and therefore some information on bubble size distribution is required for a complete understanding of the process.

Because gas holdup values vary within a reactor, average values, referring to the whole volume of the bioreactor, are usually reported. Values referring to a particular section, such as the riser or the downcomer, are much more valuable, since they provide a basis for determining liquid velocity and mixing. However, such values are less frequently reported.

The geometric design of the airlift bioreactor has a significant influence on the gas holdup. Changes in the ratio $\frac{A_d}{A_r}$, the cross-sectional areas of the downcomer and the riser, respectively, will change the liquid and gas residence time in each part of the reactor and hence their contributions to the overall holdup. Gas holdup increases with decreasing $\frac{A_d}{A_r}$, [14-17].

4.1. Gas holdup in internal airlift reactors

Correlations presented for internal-loop airlift bioreactors are shown in Table 1. These take into account liquid properties and geometric differences within a particular design. Most of the correlations take the form:

$$\varphi_r = a(J_G)^\alpha \left(\frac{A_d}{A_r}\right)^\beta (\mu_{ap})^\gamma \tag{2}$$

where φ_r is the gas holdup in the riser, J_G is the superficial gas velocity (gas volumetric flow rate per unit of cross sectional area), μ_{ap} is the effective viscosity of the liquid, and α, β, γ, and a are constants that depend on the geometry of the reactor and the properties of the liquid. The correlation can be used to predict the holdup in a system that is being designed or simulated as a function of the operating variables, the geometry of the system, or the liquid properties. Such correlations are effective for fitting data for the same type of reactor (e.g., a split-vessel reactor) with different area ratios or even different liquid viscosities, but they are mostly reactor-type specific.

The cyclic flow in the airlift bioreactor complicates the analysis of the system. The riser gas holdup depends strongly on the geometric configuration of the gas–liquid separator and the water level in the gas separator. This has been shown experimentally in a split-vessel rectangular airlift bioreactor [18], but the premise can essentially be extended to any internal loop airlift bioreactor. Analysis of the system revealed that these factors influence the gas disengagement and hence the gas recirculation in the downcomer. When this influence is taken into account and the holdup is plotted against the true gas superficial velocity, J_G, true, which is defined as the sum of the gas superficial velocity due to the freshly injected gas, Qin, and to the recirculated gas, Q_d, that is,

$$J_{G,\,true} = \left(\frac{Q_{in} + Q_d}{A_r} \right) \tag{3}$$

Then all the data for the different gas separators may be represented by a single relationship, such as equation 3. In other words, if the actual gas flow is known, the influence of gas recirculation (which depends on $\frac{A_d}{A_r}$, and the design of the gas separator) has been airlift bioreactor ready taken into account and does not need to be considered again. Nevertheless, this simple approach has a drawback in that the true gas superficial velocity is difficult to measure because the gas recirculation rate is usually not known. Thus, correlations that take into account all the variables, which may be easily measured, remain the option of choice. Table 1 shows most of the correlations of this type that have been proposed for the riser holdup in internal loop Airlift bioreactors. Comparison of a number of these correlations shows that there is reasonable agreement between the predictions of the different sources. Figure 1 can be used as an example of the actual state of-the-art in airlift bioreactor design. A number of correlations have been proposed, and three variables ($\frac{A_d}{A_r}$, l_{ap}, and J_G) have been tested by most researchers. The ranges in which these variables were studied vary from source to source.

In addition, some other variables (such as bottom clearance, top clearance or gas separator design, and surface tension) have been used by some authors but ignored by others. One example is the disengagement ratio defined by Siegel and Merchuk [19], which represents the mean horizontal path of a recirculating bubble relative to the external diameter and is equivalent to the parameter obtained by dimensional analysis [1] as:

$$M = \frac{D_S}{4D} \tag{4}$$

where D is the diameter of column and D_s the diameter of gas separator. If this parameter is not taken into account, then studies of the influence of the top clearance [13, 20] are incomplete and difficult to extrapolate to other designs.

The same can be said about the filling factor [21] given by the ratio of the gas separator volume to the total volume.

The foregoing discussion thus explains why all the correlations coincide for some ranges of these secondary variables while in other ranges they may diverge. In addition, in some cases the number of experiments may not have been sufficient to provide correlations or they may have been ill-balanced from the statistical point of view. The obvious solution to this problem lies in the collection of a large and detailed bank of reliable data that will constitute the basis for correlations with greater accuracy and validity. The safest procedure for the prediction of the gas holdup in an airlift bioreactor under design is to take data provided by researchers who have made the measurements in that particular type of reactor with the same physico-chemical properties of the system. If this option is not available, then correlation 9 in Table 1 [17], is recommended for prediction of the gas holdup in the riser.

Gas holdup in the downcomer is lower than that in the riser. The extent of this difference depends mainly on the design of the gas separator [22]. The downcomer gas holdup is linearly dependent on the riser holdup, as a consequence of the continuity of liquid flow in the reactor.

Many expressions of this type have been published [3]. At low gas flow rates, u_d is usually negligible, since most of the bubbles have enough time to disengage from the liquid in the gas separator. This usually happens at the low gas flow rates frequently used for animal cell cultures.

The gas holdup in the separator is very close to the mean gas holdup in the whole reactor [1] as long as the top clearance C_t is relatively small (one or two diameters). For larger top clearances, the behaviour of the gas separator begins to resemble that of a bubble column, and the overall performance of the reactor is influenced by this change.

In our laboratory we had some studies on the production of some secondary metabolites like phytohormones and protein hydrolysis, amino acid production, xanthophyll and new antibiotics. The next part of this chapter shows two cases were we try to show the applications of air lift bioreactors from various viewpoints, such as hydrodynamics, rheology and engineering aspects themselves. So to provide some useful engineering tools when these bioreactors takes to consideration:

5. Case 1: Hydrodynamics, mass transfer and rheological studies of Bikaverin production in an airlift bioreactor

5.1. Introduction

Antibiotics are small chemical agents (m.w. 600-800 Daltons) designed to eliminate harmful bacteria. They are produced from yeasts of fungi and bacteria. The first antibiotic discovered was penicillin in 1928/29 The first therapeutic application was in 1940 by Florey and Chain, during the Second World War when the need for antibiotics was increasing. Antibiotics belong to a group of substances called secondary metabolites. These substances appear to be unrelated to the main process of growth and reproduction. Good producers of secondary metabolites possess weak regulation of primary metabolism and visa-versa. Antibiotics are produced in limited substrate and oxygen conditions. They are produced to provide some protection against competing species in critical growth conditions. The starting points for antibiotic production are mainly amino acids and acetyl CoA with isoprene and shikimic acid also being involved. Antibiotic production commences at some point in differentiation often during sporulation. Process development consists essentially of modifying the metabolic system so that diversion of the material and biosynthesis are greatly increased. No cell growth or reproduction occurs during antibiotic production as the energy produced is being used to produce the antibiotic so that the cell can survive in the limited substrate.

After the discovery and first use of antibiotics the productivity and product concentration was increased by development of better strains or by improvement of the reaction conditions such

1	$\varphi_r = 0.44\ f_{G_r}^{0.841}\mu_{ap}^{-0.135}$ $\varphi_d = 0.297\ f_{G_r}^{0.935}$
2	$\varphi_r = 2.47\ f_{G_r}^{0.97}$
3	$\varphi_r = 0.465\ f^{0.65}\left(1 + \frac{A_d}{A_r}\right)^{-1.06}\mu_{ap}^{-0.103}$
4	$\varphi_r = 0.65\ f^{(0.603+0.078C_o)}\left(1 + \frac{A_d}{A_r}\right)^{-0.258}$ $\varphi_d = 0.46\varphi_r - 0.0244$
5	$\varphi_r = \left(0.491 - f_{G_r}^{0.706}\right)\left(\frac{A_d}{A_r}\right)^{-0.254}D_r\mu_{ap}^{-0.068}$
6	$\varphi_r = 0.16\left(\frac{J_{Gr}}{J_{1r}}\right)^{0.57}\left(1 + \frac{A_d}{A_r}\right)$ $\varphi_d = 0.79\varphi_r - 0.057$
7	$\varphi_r = 0.364\ J_{Gr}$
8	$\dfrac{\varphi_r}{1 - \varphi_r} = \dfrac{J_{G_r}^{n+2/2(n+1)}}{2^{3n+1}\left[n+2/2(n+1)\left(\frac{K}{\rho_1}\right)^{1/2(n+1)}g^{n/2(n+1)}\left(1+\frac{A_d}{A_r}\right)^{3(n+2)/4(n+1)}\right]}$
9	$\dfrac{\varphi}{(1 - \varphi)^4} = \dfrac{0.124\left(\frac{J_G\mu_1}{\sigma_1}\right)^{0.996}\left(\frac{\rho_1\sigma_1^3}{g\mu_1^4}\right)^{0.294}\frac{D_r\ 0.114}{D}}{1 - 0.276\left(1 - e^{-0.0368M_a}\right)}$
10	$\varphi_r = \dfrac{F_r}{0.415 + 4.27\left(\frac{J_{Gr} + J_{1r}}{\sqrt{gD_r}}\right)\left(\frac{g\rho_1\ D^2}{\sigma_1}\right)^{-0.188} + 1.13F_r^{1.22}M_o^{0.0386}\left(\frac{\Delta\rho}{\rho_1}\right)^{0.0386}}$
11	$\dfrac{\varphi}{(1 - \varphi)^4} = 0.16\left(\frac{J_{Gr}}{\sigma_1}\mu_1\right)M_o^{-0.283}\left(\frac{D_r}{D}\right)^{-0.222}\left(\frac{\rho_1}{\Delta\rho}\right)^{0.283} + \left(1 - 1.61\left(1 - e^{-0.00565M_a}\right)\right)$
12	$\varphi_d = 4.51\times10^6 M_o^{0.115}\left(\frac{A_r}{A_d}\right)^{4.2}\varphi_r$ When $\varphi_r < 0.0133\left(\frac{A_d}{A_r}\right)^{-1.32}$ and $\varphi_d = 0.05M_o^{-0.22}\left[\left(\frac{A_r}{A_d}\right)^{0.6}\varphi_r\right]^{0.31}M_o^{-0.0273}$ When $\varphi_r > 0.0133\left(\frac{A_d}{A_r}\right)^{-1.32}$
13	$\varphi_r = 0.0057\left[(\mu_1 - \mu_w)^{2.75} - 161\frac{73.3 - \sigma}{79.3 - \sigma}\right]\cdot f_{G_r}^{0.88}$
14	$\varphi_r = \dfrac{0.4F_r}{1 + 0.4F_r\left(1 + \frac{J_1}{J_{Gr}}\right)}$
15	$\varphi = 0.24n^{-0.6}F_r^{0.84-0.14n}G_a$

Table 1. Gas Hold-Up in Internal-Loop airlift bioreactors

as substrates and process controls. There was also a change from surface cultures to batch stirred tank reactors with complex media. The next step was the introduction of the fed batch configuration, which extended the length of the production phase and avoided repression during high substrate levels.

For maximum yield we have to design a process that only minimally involves the primary metabolism.

The main function of a properly designed bioreactor is to provide a controlled environment in order to achieve optimal growth and/or product formation. In this review we will look at the bioreactor design and the factors that are available in producing high yields of antibiotics. In general bioreactor design, the growth kinetics of the microorganism plays a key role in determining the type of reactor. Factors include; yield coefficients and maintenance require-ments, the exponential, stationary and lag phase kinetics, and the formation of growth and non-growth associated product production formation. Due to the constraints of the report, these factors are largely ignored and the main focus is on gaining knowledge on reactor configuration on the maximum yield of general antibiotics.

In our laboratory after work of ten years in research the optimal production of gibberellins we found that the *Gibberella fujkuroi* produce a potent antibiotic named Bikaverin. Bikaverin is a red pigment with specific anti-protozoal activity against *Leishmania brasiliensis* and anti-tumour activity. Additionally, bikaverin and its derivatives have a cytotoxic effect on in vitro proliferating cells of Erlich ascites carcinoma, Sarcoma 37 and leukaemia L-5178 and is a fermentation product of *Gibberella fujikuroi* or *Fusarium sp.* The formation of bikaverin precedes that of gibberellin [23] and both secondary metabolites are produced from the primary metabolite acetyl-CoA. Bikaverin is synthesized via the polyketide route while gibberellin is synthesized through the isoprenoid pathway [24].

The industrial production of these secondary metabolites is done with cultures of mycelia in liquid (submerged) or solid substrate fermentation. Models of mould growth and metabolic production based on characteristics of mycelial physiology are important to understand, design, and control those industrial fermentation processes [23, 25]. In other words these models enable us to obtain information in a practical way, facilitating fermentation analysis, and can be used to solve problems that may appear during the fermentation process.

5.2. Materials and methods

Microorganism and inoculum preparation *Gibberella fujikuroi* (Sawada) strain CDBB H-984 maintained on potato dextrose agar slants at 4 °C and sub-cultured every 2 months was used in the present work (Culture collection of the Department of Biotechnology and Bioengineer-ing, CINVESTAV-IPN, Mexico). Fully developed mycelia materials from a slant were removed by adding an isotonic solution (0.9% NaCl). The removed mycelium was used to inoculate 300 ml of fresh culture medium contained in an Erlenmeyer flask. The flask was placed in a radial shaker (200 rev min^{-1}) for 38 h at 29 ± 1°C. Subsequent to this time; the contents of the flask were used to inoculate the culture medium contained in the airlift bioreactor. The culture medium employed for the inoculum preparation is reported by Escamilla [26].

5.3. Batch culture in the airlift bioreactor

An airlift bioreactor (Applikon, Netherlands, working volume, 3.5 l) was employed in the present work (Fig 3). It consists of two concentric tubes of 4.0 and 5.0 cm of internal diameter with a settler. The air enters the bioreactor through the inner tube. A jacket filled with water allowing temperature control surrounds the bioreactor. It is also equipped with sensors of pH and dissolved oxygen to control these variables. Moreover it allows feed or retiring material from the bioreactor employing peristaltic pumps.

During the fermentation period, the pH was controlled to 3.0, temperature to 29 °C and aeration rate to 1.6 volume of air by volume of media by minute (vvm). These conditions promoted Bikaverin production with the studied strain were optimized values [26]. About 30 ml subsamples were withdrawn from the bioreactor at different times and were used to perform rheological studies. Biomass concentration was quantified by the dry weight method.

Figure 3. Airlift bioreactor used for the process for Bikaverin production

5.4. Batch culture in the airlift bioreactor

Typical culture medium contained glucose (50 g l^{-1}), NH$_4$Cl (0.75 g l^{-1}) or NH$_4$NO$_3$ (1.08 g l^{-1}), KH$_2$PO$_4$ (5 g l^{-1}), MgSO$_4$.7H$_2$O (1 g l^{-1}) and trace elements (2 ml l^{-1}). A stock solution of the trace elements used contained (g l^{-1}) 1.0 Fe SO$_4$. 7H$_2$O, 0.15 CuSO$_4$. 5H$_2$O 1.0 ZnSO$_4$. 7H$_2$O, 0.1 MnSO$_4$ 7H$_2$O, 0.1 NaMoO$_4$, 3.0 EDTA (Na$_2$ salt) 1 l of distilled water, and hydrochloric acid sufficient to clarify the solution (Barrow et al. 1960). During the fermentation period, the pH was controlled to 3.0, temperature to 29 °C and aeration rate to 1.6 v/v/m. These conditions promoted Bikaverin production with the studied strain but they are not optimized values. About 30 ml subsamples were withdrawn from the bioreactor at different times and were used to perform rheological studies. Biomass concentration was quantified by the dry weight method.

5.5. Hydrodynamics and mass transfer studies

Gas holdup was determined in the actual culture medium using an inverted U-tube manometer as described by [27]. Liquid velocities in the riser were determined measuring the time required for the liquid to travel through the riser by means of a pulse of concentrated sulphuric acid using phenolphthalein as an indicator; the same was done for the downcomer.

The mixing time was calculated as the time required obtaining a pH variation within 5% of the final pH value. For doing this, pH variation was followed after injection of a pulse of a concentrated solution of ammonium hydroxide. The volumetric mass transfer coefficient was determined employing the gassing-out method as described elsewhere [28].

5.6. Rheological studies

Rheological studies of fermentation broth were performed in a rotational rheometer (Haake, Model CV20N) equipped with a helical impeller to perform torque measurements. This type of geometry is appropriate when dealing with complex fluids and the measurement methodology is reported by Brito [29]. Rheological results, like hydrodynamics and mass transfer, are given as the average of two replicates for each sample. All the experiments were carried out in triplicate and the results that are presented are an average.

5.7. Results and discussion

5.7.1. Gas holdup

The importance of gas holdup is multifold. The gas holdup determines the residence time of the gas in the liquid and, in combination with the bubble size, influences the gas–liquid interfacial area available for mass transfer. The gas holdup impacts upon the bioreactor design because the total design volume of the bioreactor for any range of operating conditions depends on the maximum gas holdup that must be accommodated [1]. Figure 4 shows the gas holdup (ε) variation with superficial gas velocity in the riser (v_{gr}).

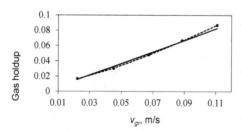

•Experimental data; — Equation 5;—Equation 14

Figure 4. Gas holdup variation with superficial gas velocity in the riser.

Experimental data were fitted to a correlation of the type of Eq. 5.

$$F = Av_{G_r}^B \qquad (5)$$

Where F could be the gas holdup (ε), the liquid velocity in the riser (v_{lr}), liquid velocity in the downcomer (v_{ld}) or the volumetric mass transfer coefficient (k_La). This type of correlation has been applied by many investigators (1, 30-33) and was derived empirically. [34] presented an analysis for Newtonian and non-Newtonian fluids where shows the theoretical basis of Eq. 5 (for the gas holdup case). He found that parameters A and B were dependent on the flow regime and on the flow behaviour index of the fluid.

Moreover, parameter A is dependent on the consistency index of the fluid, on the fluid densities and on the gravitational field. Equation 6 was obtained from fitting experimental data.

$$\varepsilon = 0.7980 \, v_{gr}^{1.0303} \qquad (6)$$

An increase in superficial gas velocity in the riser implies an increase in the quantity of gas present in the riser, that is, an increase of gas fraction in the riser [27, 32]. Chisti, [34] reports a correlation that calculates the value of B in Equation 5 (for gas holdup case). The obtained value employing this correlation is 1.2537. Gravilescu and Tudose [32] present a similar correlation, which predicts a value of 0.8434 for B. The B value obtained in the present work is between the B values obtained from these correlations that employ the flow behaviour index obtained from rheological studies. Shah [30] reported that B values in Equation 5 oscillate between 0.7 and 1.2.

5.7.2. Liquid velocity

The liquid circulation in airlift bioreactors originates from the difference in bulk densities of the fluids in the riser and the downcomer. The liquid velocity, while itself controlled by the gas holdups in the riser and the downcomer, in turn affects these holdups by either enhancing or reducing the velocity of bubble rise. In addition, liquid velocity affects turbulence, the fluid-reactor wall heat transfer coefficients, the gas-liquid mass transfer and the shear forces to which the microorganism are exposed. Figure 5 shows liquid velocities variation in the riser and the downcomer as a function of superficial gas velocity in the riser.

Liquid velocities in the riser (v_{lr}) and in the downcomer (v_{ld}) were fitted to correlations of the type of Equation 1 and Equations 7 and 8 were obtained.

$$v_{lr} = 1.3335 \, v_{gr}^{0.3503} \qquad (7)$$

$$v_{ld} = 0.8716 \, v_{gr}^{0.2970} \qquad (8)$$

Freitas and Teixeira [35] point out that B value in equation 7 must be close to 0.3333 for the liquid velocity in the riser since this value was theoretically derived by Kawase [36] and others. The B value obtained in the present work (0.3503) is only a little bit higher than 0.3333. Freitas

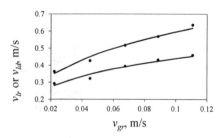

•Experimental data — Equation 7 or 8

Figure 5. Liquid velocities as a function of superficial gas velocity in the riser.

and Teixeira [35] also obtain that B value for the liquid velocity in the downcomer is less than B value for the liquid velocity in the riser which agrees with the results obtained in this work. Liquid velocities in the riser and in the downcomer increase with an increase in gas velocity in the riser due to an increase in the density difference of the fluids in the riser and the downcomer.

5.7.3. Mixing time

Mixing in airlift bioreactors may be considered to have two contributing components: backmixing due to recirculation and axial dispersion in the riser and downcomer due to turbulence and differential velocities of the gas and liquid phases [37](Ch. Mixing time is used as a basis for comparing various reactors as well as a parameter for scaling up [32]. Figure 6 shows the mixing time variation with the superficial gas velocity in the riser.

Once again, the mixing time variation was fitted to a correlation of the type of Equation 5 and Equation 8 was obtained.

$$t_m = 5.0684 \, v_{gr}^{-0.3628} \qquad\qquad (9)$$

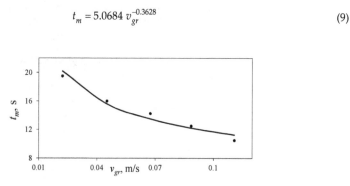

• Experimental data — Equation 5

Figure 6. Mixing time as a function of superficial gas velocity in the riser.

Choi *et al.*, [37] report a *B* value in Equation 5 of –0.36 while Freitas and Teixeira [35] report a *B* value equal to –0.417. The *B* value obtained in this work is similar to the value reported by Choi *et al.*, [37]. The mixing time decreases with an increase in superficial gas velocity in the riser since the fluid moves more often to the degassing zone where most of the mixing phenomenon takes place due to the ring vortices formed above the draught tube [35].

5.7.4. Volumetric mass transfer coefficient

One of the major reasons that oxygen transfer can play an important role in many biological processes is certainly the limited oxygen capacity of the fermentation broth due to the low solubility of oxygen. The volumetric mass transfer coefficient ($k_L a$) is the parameter that characterizes gas-liquid oxygen transfer in bioreactors. One of the commonest employed scale-up criteria is constant $k_L a$. The influences of various design (i.e., bioreactor type and geometry), system (i.e., fluid properties) and operation (i.e., liquid and gas velocities) variables on $k_L a$ must be evaluated so that design and operation are carried out to optimize $k_L a$ [37].

The value of the volumetric mass transfer coefficient determined for a microbial system can differ substantially from those obtained for the oxygen absorption in water or in simple aqueous solutions, i.e., in static systems with an invariable composition of the liquid media along the time. Hence $k_L a$ should be determined in bioreactors which involve the actual media and microbial population [38]. Figure 7 shows the volumetric mass transfer coefficient variation with the superficial gas velocity in the riser.

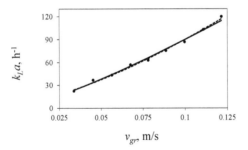

• Experimental data — Equation 6 — Equation 12

Figure 7. Effect of the superficial gas velocity in the riser on $k_L a$.

Experimental data shown in Figure 7 were fitted to a correlation of the type of Equation 1 and Equation 6 was obtained.

$$k_L a = 0.4337 \, v_{gr}^{1.2398} \tag{10}$$

[39] report a *B* value in Equation 6 equal to 1.33 and Schügerl *et al.*, [40] report a value of 1.58. The value of 1.2398, obtained in this work, is closed to these last values.

Volumetric mass transfer coefficient (k_La) increases with an increase in superficial gas velocity in the riser due to an increase in gas holdup which increases the available area for oxygen transfer. Moreover an increase in the superficial gas velocity in the riser increases the liquid velocity which decreases the thickness of the gas-liquid boundary layer decreasing the mass transfer resistance.

Figure 8 shows the evolution of k_La through fermentation course employing two different nitrogen sources. The k_La decreases in the first hours of fermentation and reaches a minimum value at about 24 hours. After this time the k_La starts to increase and after 48 hours of fermentation it reaches a more or less constant value which remains till the end of fermentation process. This behaviour is similar irrespective of the nitrogen source and will be discussed with the rheological results evidence.

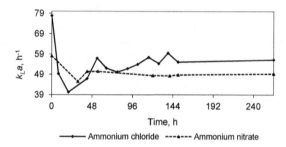

Figure 8. Sthe evolution of k_La through fermentation course employing two different nitrogen sources.

Figure 9 shows the relation between gas holdup and k_La. Mc Manamey and Wase [21] point out that the volumetric mass transfer coefficient is dependent on gas holdup in pneumatically agitated systems. The later was experimentally determined in bubble columns by Akita and Yoshida and other authors [30, 41, 42] they mention that this was expectable since both the volumetric mass transfer coefficient and the holdup present similar correlations with the superficial gas velocity. Mc Manamey and Wase [21] proposed a correlation similar to Equation 1 to relate volumetric mass transfer coefficient with gas holdup. Equation 7 presents the obtained result.

$$k_La = 0.2883 \; \varepsilon^{0.9562} \tag{11}$$

Akita and Yoshida [41] and Prokop *et al.* [42] found that the exponent in Equation 7 oscillates between 0.8 and 1.1.

It is well known [16] that logarithmic scale plots of k_La vs. $\varepsilon/(1-\varepsilon)$ for any particular data set should have a unit slope according to Equation 8.

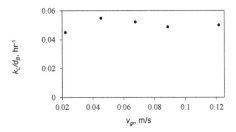

Figure 10. The k_L/d_B ratio as a function of superficial gas velocity.

$$\ln k_L a = \ln\left(6\frac{k_L}{d_B}\right) + \ln\frac{\varepsilon}{(1-\varepsilon)} \tag{12}$$

Where k_L is the mass transfer coefficient and d_B is the bubble diameter. Even though the latter is a generally known fact, few investigators determined these slopes for their data to ascertain the validity of their experimental results. Figure 10 shows this analysis for the experimental data of the present work obtaining a slope of 1.034. Chisti [16] shows the same analysis for two different data set and obtained slopes of 1.020 and 1.056.

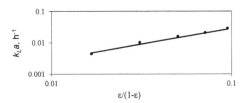

Figure 9. $k_L a$ vs. gas holdup in the airlift bioreactor, unit slope.

A rearrangement of Equation 8 leads to Equation 9 which results are shown in Figure 10

$$\frac{k_L}{d_B} = \frac{k_L a(1-\varepsilon)}{6\varepsilon} \tag{13}$$

The average value of k_L/d_B obtained in the present work is 0.050 s^{-1}. Chisti [27] performed a similar analysis for 97 data points obtained from several different reactors and found an average value of 0.053 s^{-1}. The foregoing observations have important scale-up implications. In large industrial fermenters the $k_L a$ determination is not only difficult, but there is uncertainty as to whether the measured results reflect the real $k_L a$ or not. The gas holdup measurements

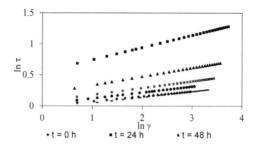

Figure 11. Typical rheogram employing impeller viscometer.

on these reactors are relatively easy to carry out, however. Thus, Equation 9 can help to estimate $k_L a$ in these reactors once holdup measurements have been made [1].

5.7.5. Rheology

Rheological parameters such as the flow index (n) and the consistency index (K) depend on such factors as the concentration of solids in the broth, the morphology (length, diameter, degree of branching, shape) of the particles, the growth conditions (flexibility of cell wall and particle), the microbial species and the osmotic pressure of the suspending liquid, among others possible factors.

For the case of mycelia cultures, as the biomass concentration increases the broth becomes more viscous and non-Newtonian; leading to substantial decreases in oxygen transfer rates. This effect is often important since for many aerobic processes involving viscous non-Newtonian broths oxygen supply is the limiting factor determining bioreactor productivity [43]. Apparent viscosity is a widely used design parameter which correlates mass transfer and hydrodynamic parameters for viscous non-Newtonian systems [44].

It is worth to mention that the present work uses impeller viscometer for performing rheological studies avoiding the use of other geometries, i.e., concentric tubes or cone and plate, overcoming associated problems with these geometries such sedimentation, solids compacting and jamming between measuring surfaces or pellet destruction [45].

Rheograms obtained for the fermentations employing different nitrogen source were fitted to Ostwald-de Waele model (power law) and in both cases a pseudoplastic behaviour for the fermentative medium was found. Figure 11 shows the results of consistency and flow indexes for these two fermentations where similar results were obtained.

During the first 24 hours of fermentation, medium viscosity increases due to exponential growth of mycelia (no lag phase is present) which causes a $k_L a$ decrease in Figure 9. After this time, the formation of pellets by the fungus starts to occur reflected in a decrease of medium viscosity and hence an increase in $k_L a$ value in Figure 10. After 72 hours of fermentation the medium viscosity was practically unchanged because the stationary growth phase is reached by the fungus reflected in practically constant values of medium viscosity and $k_L a$. With the

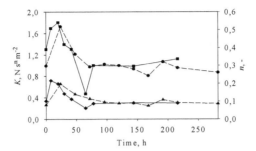

Figure 12. K and n through fermentation time in the airlift bioreactor. • K for ammonium nitrate ▲ n for ammonium nitrate ■ K for ammonium chloride ♦ n for ammonium chloride,

aid of rheological studies is possible to use correlations of the type of Equation 10 to relate holdup and volumetric mass transfer coefficient with fermentation medium viscosity [20, 31, 39, 44] to obtain Equations 12 and 13.

$$F = A v_{gr}^{B} \mu_{app}^{C} \tag{14}$$

$$k_L a = 0.0036 \, v_{gr}^{0.3775} \mu_{app}^{-0.5488} \tag{15}$$

$$\varepsilon = 0.0072 \, v_{gr}^{0.2381} \mu_{app}^{-0.5703} \tag{16}$$

Figures 2 and 6 show experimental data fitting for holdup and $k_L a$, respectively. As it was expectable, Equations 12 and 13 present a better fit to experimental data than that obtained with the aid of Equations 2 and 3 due to the existence of an extra adjustable parameter.

As can be seen in Fig. 13, there is no lag phase and exponential growth of mycelia starts immediately and ceases during the first 24 h of fermentation. The later causes the medium viscosity to increase (K and n increase in Fig. 12), which causes a $k_L a$ decrease in Fig. 6. After 24 h of fermentation, the formation of pellets by the fungus starts to occur, reflected in a decrease of medium viscosity (K and n start to decrease in Fig. 12) and hence an increase in $k_L a$ value in Fig. 5. After 72 h of fermentation the medium viscosity was practically unchanged (K and n remain constant in Fig. 12) because the stationary growth phase is reached by the fungus reflected in practically constant values of medium viscosity and $k_L a$. Also, after 72 h of fermentation, the pellet formation process by the fungus stops.

Figure 13 shows the correlation between consistency and flow indexes with biomass concentration. Experimental data were fitted to Eqs. 12 and 13 proposed in the present work. Optimized values for constants in Eqs. 12 and 13 are summarized in Table 1.

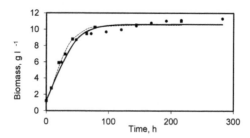

Figure 13. Growth kinetics employing ammonium chloride (♦) or ammonium nitrate (•) as nitrogen source.

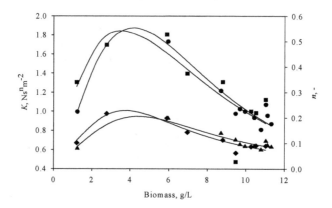

Figure 14. K and n as a function of biomass concentration in the airlift bioreactor. • K for ammonium nitrate ▲ n for ammonium nitrate ■ K for ammonium chloride ♦ n for ammonium chloride

$$K = \frac{c_1}{\left(1 + \frac{c_2}{x}\right) + \left(\frac{x}{c_3}\right)^2}$$ (17)

$$n = \frac{c_1}{\left(1 + \frac{c_2}{x}\right) + \left(\frac{x}{c_3}\right)^2}$$ (18)

With the aid of rheological studies is possible to use correlations of the type of Equation 14 to relate gas holdup and volumetric mass transfer coefficient with fermentation medium viscosity [20, 31, 39, 44] to obtain Equations 15 and 16.

$$F = A v_{gr}^{B} \mu_{app}^{C} \qquad (19)$$

$$k_L a = 0.0036 \ v_{gr}^{0.3775} \mu_{app}^{-0.5488} \qquad (20)$$

$$\varepsilon = 0.0072 \ v_{gr}^{0.2381} \mu_{app}^{-0.5703} \qquad (21)$$

Figures 2 and 5 show experimental data fitting for gas holdup and $k_L a$, respectively. As it was expectable, Eqs. 15 and 16 present a better fit to experimental data than that obtained with the aid of Eqs. 2 and 3 due to the existence of an extra adjustable parameter.

5.8. Conclusions

In the present work preliminary hydrodynamics, mass transfer and rheological studies of Bikaverin production in an airlift bioreactor were achieved and basic correlations between gas holdup, liquid velocity in the riser, liquid velocity in the downcomer, mixing time and volumetric mass transfer coefficient with superficial gas velocity in the riser were obtained.

Adjustable parameters calculated for each variable were compared with literature reported values and a good agreement was obtained.

The gassing out method was successfully applied in determining volumetric mass transfer through fermentation time employing two different nitrogen sources. Irrespective of the nitrogen source the volumetric mass transfer behaviour was similar and it was explained in terms of the fungus growth and changes in its morphology, which affect the culture medium rheology.

Pellet formation by the fungus was used to explain the increase of $k_L a$ or the decrease of medium viscosity. In both fermentations, $k_L a$ decreases as exponential growth of the fungus occurs and reaches an asymptotic value once the stationary growth phase is reached. A helical impeller was employed successfully for rheological studies, avoiding problems of settling, jamming or pellet destruction, finding that the culture medium behaves as a pseudoplastic fluid. Rheological measurements were used to correlate gas holdup and $k_L a$ with apparent culture medium viscosity. Once again, for both fermentations, apparent viscosity increases as exponential growth of the fungus occurs and reaches an asymptotic value once the stationary growth phase is reached.

A satisfactory validation of experimental data for gas holdup and volumetric mass transfer coefficient was performed which allows the employment of these data in scale-up strategies.

6. Case 2: Studies on the kinetics, oxygen mass transfer and rheology in the l-lysine prodution by *Corynebaterium glutamicum*

6.1. Introduction

The industrial application of amino acids in broiler feed has a long history, from the late 1950's have been used to increase the efficiency of the food they eat these animals. Lysine is one of these amino acids to its importance as a feed additive for pigs and poultry, is that it increases the willingness of proteins, bone growth, ossification and stimulates cell division.

Lysine used as an additive for food, is imported because lysine production nationally not exist. The approximate amount of lysine produced worldwide is 550 000 tons per year and almost everything is produced by international companies. Only in the state of Guanajuato, the estimated demand of 300 tons of lysine [46], so that the implementation of appropriate technology for the production of this important amino acid, reduce production costs of animal feed feedlot to reduce imports and generate jobs in the country [47].

Lysine can be produced by chemical synthesis or by enzymatic or microbiological processes. Lysine for obtaining a chemical synthesis is expensive and inefficient process that also by this method are obtained racemic mixtures of D and L forms must then be processed to obtain the L-form which is biologically active. Microbiological fermentation processes are more efficient and direct methods to be based on the accumulation of amino acid that is excreted by the organism in culture media and / or fermentation containing high sugar concentrations and ammonium ions at neutral pH and under aerobic conditions in crops batches [48].

The most commonly used species for the production of lysine is the *Corynebaterium glutamicum* are employed although *Arthrobacter, Brevbibacterium, Microbacterium and Micrococuus*.

Traditionally, the production process of lysine by microbiological fermentation is carried out in stirred vessels, the characteristics of this type of reactor, sometimes, it is economically inconvenient. A better alternative to this configuration are the airlift reactors, the advantages are: low shear, high-speed transfer of oxygen and good mixing, implying a better mass transfer eliminating concentration gradients of either the medium components, avoiding sedimentation of the cells, thereby creating a more favourable environment for development and maintenance, increasing yields and production of lysine.

6.2. Methodology

Inoculum development. Inoculated to the reactor biomass was obtained from a pre-inoculum of ATCC 21253 in lyophilized strain *Corynebacterium glutamicum*.

Experiments in the bioreactor. The batch fermentations were carried out in a reactor airlift with a working volume of 3.4 litres at 30° C (Fig 15). Dissolved oxygen was monitored with a polarographic electrode. The pH of the culture pH was controlled at 7.0 by addition of a solution of 70% NH_4OH. The bioreactor has a condenser which minimizes the error by evaporation losses. The quantity of biomass was inoculated approximately 10% of the bioreactor working volume and sampled periodically.

Figure 15. Airlift bioreactor for L-Lysine production

6.3. Mass transfer

For the determination of K_La was used the gas elimination technique [49] and dynamic technique [28]. Measurements for the calculation of K_La by both methods were carried out at 30 ° C and pH 7.0, with a working volume of 3.4 litres of sterile fermentative medium.

Evaluation of the parameters α, β and m_s, the equations were solved using the Runge-Kutta 4th order and an adjustment of the experimental data by nonlinear regression using the GREG program.

The growth rate model is as follows

$$\frac{dX}{dt} = \mu X \left(1 - \frac{X}{L}\right) \tag{22}$$

Where μ is the specific growth rate and L is the maximum value that people can achieve.

The model of product formation rate where the rate of formation is related to the rate of growth:

$$\frac{dP}{dt} = \alpha \frac{dX}{dt} \tag{23}$$

Where α is a constant stoichiometric. In the case where the product is formed independently of the speed of growth:

$$\frac{dP}{dt} = \beta X \tag{24}$$

where β is a proportionality constant. The constant β is similar to the enzyme activity [28]

The substrate consumption model is represented by the following equation:

$$r_S = \frac{r_X}{Y_{X/S}'} + \frac{r_P}{Y_{P/S}'} + m_S X \tag{25}$$

6.4. Results and discussion

The results obtained with the gas elimination technique, allowed calculating the maximum oxygen transfer rate in the system, the dissolved oxygen conditions required.

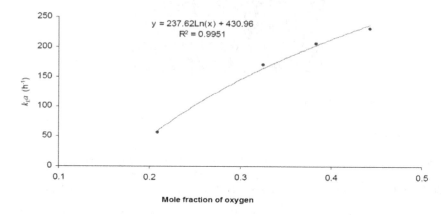

Figure 16. Adjusting K_La experimental values obtained by the technique of gas removal.

Finding no correlations suitable K_La experimental data obtained by the technique of gas phase were adjusted to a logarithmic trend line shown in Figure 16.

Figure 17, shows that the production of lysine started between 11 and 12 hours. From 21 up to 46 h the lysine production rate was kept constant (0.385 g / l of lysine h) and declined at 53 h the reduction in production rate was 22% approximately. The overall yield Y_P / S was 0.244 g of lysine / g glucose (at 53 h Y_P / S = 0.223 g lysine / g glucose).

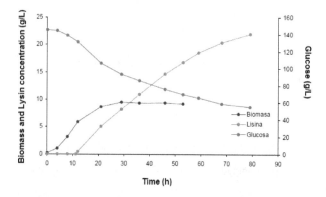

Figure 17. Kinetics of growth, production and consumption with initial glucose 145 g / l.

Parameter	102g/l glucose	145g/l glucose
μ	0.422	0.362
L	11.36	9.26
α	0.974	0.72
β	0.0405	0.0378
Y'X/S	0.42	0.38
Y'P/S	0.36	0.36
ms	–	0.0245

Table 2. Models parameter of growth, production, and lysine and glucose consumption.

In the experiment with 102 g / l initial glucose (Figure 17), lysine production started between 9 and 10 hours. From 22 up to 46 hours the lysine production rate was kept constant (0.475 g / l h of lysine), at 52 hours the reduction in production rate was only 8%. The overall yield Y_P / S was 0.247 g of lysine / g glucose.

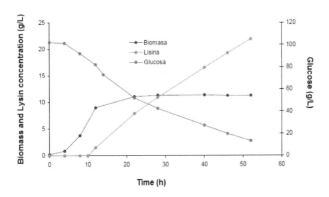

Figure 18. Kinetics of growth, production and consumption with initial glucose 102 g / l.

The model parameters of growth, production and consumption presented in equations [1] to [4] are presented in Table 2.

For evaluation of the parameters α, β and m_s. The equations were solved using the Runge-Kutta 4th order and an adjustment of the experimental data by nonlinear regression using the GREG program [50].

In evaluating the parameters α, β and m_s, the other parameters were kept constant and adjusting the experimental data was performed with the data obtained from 12 h until the end of the maintenance phase (phase to production rate constant).

In Figures 19 and 20 shows the experimental data and the values generated by equations (solid lines) and can be seen that models correctly predict the growth of the microorganism, product formation and substrate consumption from the beginning of the lysine production phase, until the end of the maintenance phase.

Initial Glucose (g/L)	$L - X_0$	μ	$Y_{X/Thr}$	q_{Thr}
145	9.04	0.362	30.1	0.0120
102	11.14	0.422	37.1	0.0114

Table 3. Results obtained from the specified speed of consumption of threonine to two different initial concentrations of glucose.

The rate of consumption of threonine is affected by the initial concentration of glucose due to the effect that the osmotic pressure produced in the cells.

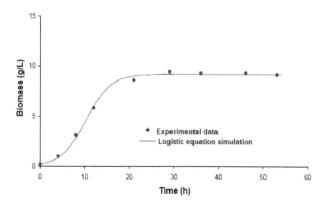

Figure 19. Setting the biomass data to the logistic model (glucose = 145 g/L)

Based on the data in Table 4, an experiment was conducted with airflow of 1 vvm and initial concentration of 150 g / l and 0.6 g / l glucose and L-threonine respectively (Figure 20). The dissolved oxygen began to decrease at approximately 2 h and reached a level below 5% saturation at approximately 20 hours. In Figure 20 it can be seen the effect of oxygen in the growth rate, until 8 h, $\mu = 0.250\ s^{-1}$ with a saturation percentage of dissolved oxygen of 20% and 22 to 46, $\mu = 0.0135$, after the culture was oxygen limited.

Air flow (vvm)	ml of antifoam uptake (12 h)	liquid entrained by the air trapped in the condenser
0.5	<1	-
1.0	15	2.2
1.5	27.5	4.4
2.0	55.5	8.0
2.5	>75	11.2
3.0	-	16.8

Table 4. Exploratory experiments for found the optimal initial air flow in the airlift bioreactor.

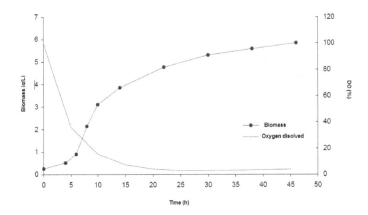

Figure 20. Oxygen limiting effect on the biomass formation.

6.5. Experimental determination of the volumetric coefficient of oxygen transfer

Determination of the solubility of oxygen in the fermentative culture. Because in the experiments with variable air flow was not possible to maintain the dissolved oxygen concentration required for the fermentation experiments was determined by enriching the airflow with oxygen. Were tested for solubility of oxygen in the fermentative culture with various concentrations of glucose, 100, 140 and 180 g / l were obtained the following results:

Glucose (g/l)	DO (%)	MgO_2/l
180	85.22	5.27
140	85.98	5.31
100	87.30	5.40

Table 5. Solubility of oxygen in the fermentative medium with different glucose concentrations.

The difference in concentration of dissolved oxygen concentration between the highest and lowest blood glucose was only 2.11% [51]. Therefore experiments to determine $k_L a$ is conducted in fermentative medium with 140 g / l glucose. One way to improve oxygen transfer in the fermentations is to increase the solubility by increasing the mole fraction of gas in the airflow. Based on this, we measured the concentration of dissolved oxygen in the bioreactor by varying the mole fraction of oxygen in the air flow. The results are presented in Table 6.

Oxygen flow rate (vvm)	Air flow rate (vvm)	Oxygen molar fraction	mmol O_2/l	mg O_2/l
0	3400	0.209	0.165	5.27
500	2900	0.325	0.255	8.15
750	2650	0.383	0.307	9.81
1000	2400	0.442	0.349	11.18
1500	1900	0.558	0.441	14.10

Table 6. Experimental results of the measurement of dissolved oxygen.

Experimental data of solubility of oxygen, according to Henry's Law can be adjusted to a straight line (Figure 21). These data were used in subsequent calculations.

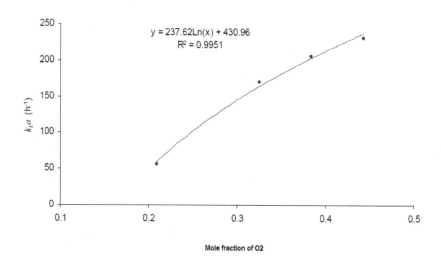

Figure 21. Adjusting $k_L a$ experimental values obtained by the technique of gas removal.

In 2003, Ensari and Lim [52] proposed a model in which incorporated the rate of oxygen consumption to describe the kinetics of fermentation of L-lysine and handled 0,266 mmolO$_2$ /

l as 100% saturation, 30% of this amount is equivalent mgO$_2$ to 2.55 / l. If we take the value of 2.55 mgO$_2$ / l, as the minimum value of dissolved oxygen should be maintained during the fermentation and used to calculate the oxygen transfer rate, as the mole fraction of oxygen increases in the air flow, the percentage of dissolved oxygen is reduced, due to the increase in the concentration gradient, which is the driving force for oxygen transfer.

Once the solubility data obtained at different levels of enrichment, we proceeded to calculate the values of the volumetric coefficient of oxygen transfer. K$_L$a value is proportional to the increase in the molar fraction of oxygen in the air flow, this is because increasing the amount of oxygen in the flow, and the contact area is higher. The results obtained with the gas elimination technique, are shown in Table 7 and were useful because together with the values of solubility of oxygen, were used to calculate the maximum oxygen transfer rate in the system, the conditions of dissolved oxygen required.

Mole fraction of oxygen	k$_L$a (h^{-1})	Number of adjusted data	R^2
0.209	57	10	0.9977
0.209	56	10	0.9979
0.325	171	5	0.9958
0.325	169	5	0.9903
0.383	205	5	0.9921
0.383	208	5	0.9911
0.442	233	3	0.999
0.442	230	3	0.9975

Table 7. Experimental K$_L$a data obtained by the technique of gassing out.

While gassing out method showed a good correlation decided to try the dynamic method to see if you got a better approximation.

6.6. Dynamic method

In Figure 22, shows the data obtained during the development of direct measurements were performed to calculate the volumetric coefficient of oxygen transfer. Data from the first phase of the experiment were used to measure the oxygen consumption rate in the second phase, the coefficient of volumetric oxygen transfer. In fermentation, the oxygen transfer rate can be calculated using the following equation:

$$\frac{dC_L}{dt} = K_L\, a\left(C_L^* - C_L\right) - Q_{O_2}X \tag{26}$$

Where, Q$_{O2}$ is the specific rate of oxygen consumption, which can be defined as the specific growth rate between the yields of oxygen.

$$Q_{O_2} = \frac{\mu}{Y_{O_2}} \tag{27}$$

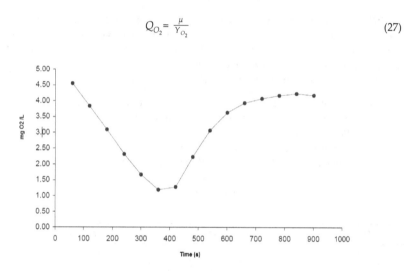

Figure 22. Typical response during the development of the dynamic technique of oxygen uptake

Q_{O2} value can be considered constant during the exponential phase of growth, since, during this stage, both the specific growth rate and yield of oxygen remained constant [25].

This dependency has been used by several authors [51, 53] for correlating the concentration of biomass and the rate of oxygen consumption by a constant parameter, at least during the exponential growth phase. For these experiments, we used a concentration we used a concentration of 140 g / l glucose and the flow of air not enriched with oxygen. The results of calculating the specific rate of oxygen consumption are shown in Table 8. The average value of Q_{O2}, was 159 mg O_2 / g of cells per h, with a standard deviation of 1,414. With this information it is possible to determine the amount of oxygen required for a given biomass concentration in a fermentation and establish the amount of oxygen to be provided shall in the airflow.

Molar fraction molar of Oxygen	$Q_{o2}X$	R^2	X (g cells)	Q_{o2} (mgO$_2$/g cell h)
0.209	0.01242	0.999	0.28	160
0.209	0.01443	0.997	0.33	158

Table 8. Calculation of the specific rate of oxygen consumption

To calculate the volumetric coefficient of oxygen transfer, we generated a graph (Figure 23) of $dC / dt + Q_{o2}X$ against C_L, where the slope of the line is $-1/k_La$ and intercept and C_L. k_La value of the average between the two experiments was 53 with a standard deviation of 1.414 that show a good acceptation.

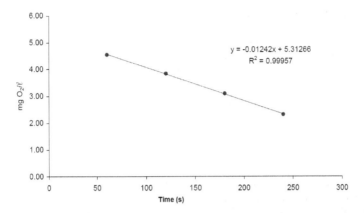

Figure 23. Calculation of the oxygen consumption rate using a direct method, the slope of the line is equal to $-Q_{O_2}X$.

Comparing the values obtained with the technique $k_L a$ gas removal and direct measurement shows that the $k_L a$ value measured in the early hours of the fermentation is very similar to the calculated half fermentative bacteria free. Therefore, to calculate the oxygen transfer rate was determined considering constant since the viscosity during the fermentation does not vary significantly. The solubility of oxygen, may be affected if, therefore, the dissolved oxygen electrode was calibrated at the start of fermentation to a high saturation conditions chosen and the operation of the fermenter was held at the level of dissolved oxygen concentration corresponding.

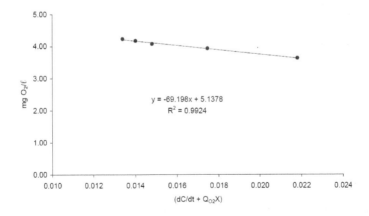

Figure 24. Calculating the volumetric coefficient of oxygen transfer, using a direct method, the slope of the line is equal to $-1/k_L a$.

6.7. Oxygen enrichment experiments

In the case of the 21253 strain of *Corynebacterium glutamicum*, the course of the airlift fermentation in the bioreactor can be divided in 4 phases. The first phase is characterized by exponential growth of the organism (as long as there are no restrictions of any nutrient). The duration of this phase depends on the initial concentration of threonine in the fermentative medium. The depletion of threonine, the second phase begins. At this point lysine production begins, the dissolved oxygen concentration is increased (this implies a decrease in oxygen consumption at the maximum consumption of oxygen is given up to the point of exhaustion), the cell concentration continues to increase and eventually reaches a maximum of approximately 1.6 times the amount shown at the point of exhaustion of threonine 1.7 times [54].

The end of biomass production is due to the depletion of threonine [48] and begins the third phase, in which occurs the maintenance stage and the lysine production rate remains constant. In the fourth phase, the decrease in the production rate is remarkable, because it is not possible cell turnover by the lack of threonine, leucine and methionine, therefore, the biomass concentration decreases.

The yield of threonine should vary depending on the initial concentration of glucose due to growth inhibition that occurs in high concentrations. So far, in experiments with variable air flow, with an initial concentration average of 145 g / l of glucose values were $\mu = 0.331$ (h^{-1}) $Y_X / T_{hr} = 23.10$ (g biomass / g of threonine) and $q_{Thr} = 0.0143$ (g threonine / g biomass h) in the bioreactor airlift was obtained in a yield of 40.50 (g biomass / g of threonine). In both cases the oxygen limited the growth.

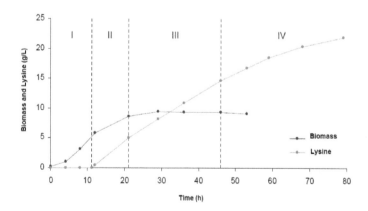

Figure 25. Stages in the lysine fermentation with *Corynebacterium glutamicum* ATCC 21253 in an airlift bioreactor.

To determine the parameters of the model without oxygen limitation were conducted two experiments with initial concentrations of 145 and 100 g / l glucose and 0.3 g / l of threonine. For fermentation with 145 g / l initial glucose, the maximum molar fraction of oxygen was 0.325 and 0.349 for fermentation with 102 g / l initial glucose. In Table 8 shows the final results of the two fermentations. The kinetics of growth, product formation and substrate consumption can be observed in Figures 26 and 27.

Initial Glucose (g/l)	Biomass (g/l)	Lysine (g/l)	Residual glucose (g/l)	Process time(h)
145	9.26	21.96	55.0	79
102	11.36	21.9	13.2	52

Table 9. Final results of the fermentations with oxygen-enriched air at the same culture conditions.

The yield $Y_{X/Thr}$ between both experiments varied due to the inhibition caused by the initial concentration of glucose, for the experiment with 145 g / l initial glucose Y_X / T_{hr} = 30.13 (g biomass / g of threonine) and in which was used 102 g / l initial glucose was 37.1 (g biomass / g of threonine). This indicates that the amount of threonine to increase relative to the amount of glucose for an adequate quantity of biomass in each fermentation and minimize residual glucose concentrations.

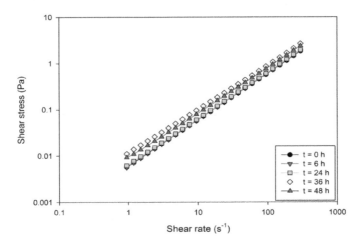

Figure 26. Rheological behaviour when the shear stress versus the shear stress change respect the times and with initial glucose concentration of 100 g / L.

6.8. Results of dynamic rheological behaviour of fermentation process

The model that best described the shear stress in function of the shear rate was

Biofluido behaviour in initial concentrations of 100 (Exp. 1), 140 (Exp. 2) and 180 g / l (Exp. 3) glucose and 1 vvm of air flow, presents different behaviour with respect to its apparent viscosity. Analyses were performed on a controlled stress rheometer with ARG2 type concentric cylinder geometry and with an observation range of 0.1 to 300 rps. The following figures

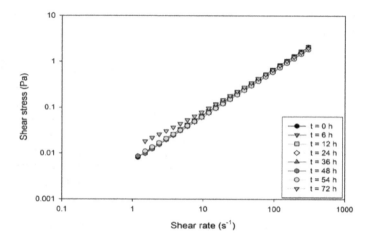

Figure 27. Rheological behaviour when the shear stress versus the shear stress change respect the times and with initial glucose concentration of 140 g / L.

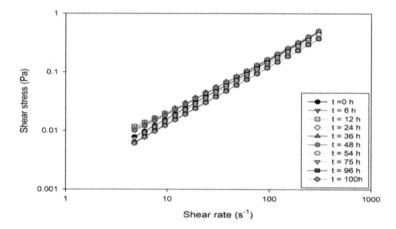

Figure 28. Rheological behaviour when the shear stress versus the shear stress change respect the times and with initial glucose concentration of 180 g / L.

show the behaviour of the shear stress on the shear rate and at different times to take sample of fermentation broth.

The rheological analysis was carried out both in increasing the shear rate and decreasing the same and no significant change was observed i.e. a curve passes over another, this being a feature of pseudoplastic fluids.

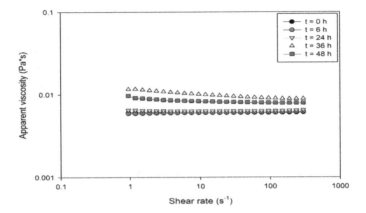

Figure 29. Apparent viscosity versus Shear rate at different times, with initial glucose 100 g / l.

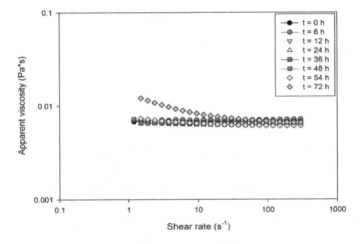

Figure 30. Apparent viscosity versus Shear rate at different times, with initial glucose 140 g / l.

It is evident that the maximum apparent viscosity is obtained between 24 and 36 hours the fermentation process, and subtly observed as the shear stress decreases with increasing the initial concentration of glucose in the medium (Figure 28) for the same shear rate.

The following graphs we can provide additional information to what occurs with respect to the apparent viscosity of the medium and in the same manner at different initial values of glucose.

Similarly one can conclude that the reduced carbon source leads to a decrease in apparent viscosity. The rheological parameters can be described in terms of the model of power law

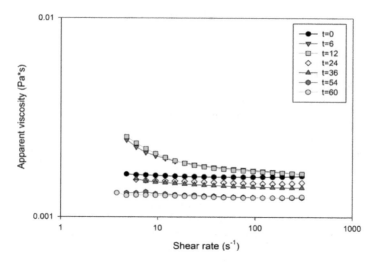

Figure 31. Apparent viscosity versus Shear rate at different times, with initial glucose 180 g / l.

fluids for pseudoplastic, which show a nonlinear relationship between shear stress (τ) and shear rate (γ).

$$\tau = k^* \gamma^n \tag{28}$$

The constant k is a measure of the consistency of the fluid consistency index is called, and the exponent n is indicative of the deviation of the fluid flow about the behaviour and often called Newtonian behaviour index. For pseudoplastic fluids it holds that n <1, while n> 1 means a dilatant flow behaviour. The power law representing the Newtonian fluid when n = 1. To look more closely shown in the following tables the evolution of the flow rate and consistency index with respect to time and the initial glucose concentration.

Time (h)	K (Pa*s)	n
0	0.00509	1.0055
6	0.00580	1.0111
24	0.00599	1.0136
36	0.01010	0.9779
48	0.00827	0.9926

Table 10. Evolution of the flow and consistency index when were used the initial glucose 100 g / l.

Time (h)	K (Pa*s)	n
0	0.00640	1.0141
6	0.00682	1.0093
12	0.00320	1.0133
24	0.00625	1.0095
36	0.00620	1.0033
48	0.00626	1.0048
54	0.00638	0.9929
72	0.00864	0.9514

Table 11. Evolution of the flow and consistency index when were used the initial glucose 140 g / l.

Time (h)	K (Pa*s)	n
0	0.00157	1.0028
6	0.00209	0.9577
12	0.00202	0.9633
24	0.00227	0.9389
36	0.00212	0.9589
48	0.00199	0.9735
54	0.00152	0.9944
72	0.00149	0.9891
96	0.00133	0.9889
102	0.00126	0.9995

Table 12. Evolution of the flow and consistency indices when were used the initial glucose 180 g / l.

With these data it is easy to see how the cross breeding ground of a pseudoplastic to dilatant, if however it remains very close to the threshold of Newtonian behaviour.

Trying to interpret which may be the kinetic relationship with the following graphs are presented, which illustrates the behaviour of the indices of both flow and consistency.

Can be seen from the graph that during the growth phase of the microorganism fluid behaves as pseudoplastic and when it reaches the stationary phase changes dilatant. The interesting thing is that presented in this final stage the culture broth is such that can be separated easily from the microorganism and undesirable solids, providing the following extraction step of the lysine.

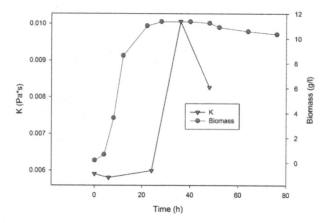

Figure 32. Behaviour consistency index (k) and biomass through fermentation time, using an initial glucose concentration of 100 g / l.

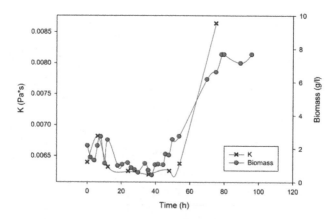

Figure 33. Behaviour consistency index (k) and biomass through fermentation time, using an initial glucose concentration of 140 g / l.

Shows the effect of glucose on the growth of the microorganism, presenting a catabolic repression, and which manifests itself in changing the rheology of the system. We can see that it is present a phase lag increased with increasing initial glucose concentration and the fluid is Newtonian behaviour.

When using the higher initial glucose concentration shows a higher apparent viscosity and thus shows the suppressive effect on growth of the microorganism, once decreases the concentration of glucose over time changes the rheological behaviour of the system, but with a low lysine production.

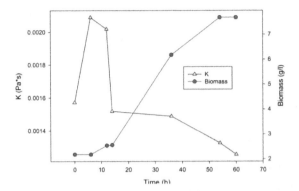

Figure 34. Behaviour consistency index (k) and biomass through fermentation time, using an initial glucose concentration of 180 g / l.

With this new information becomes more evident that the media has a high apparent viscosity at about 24 hours of fermentation, if this information is compared by the kinetics osmolality can be concluded to be due to agglomeration of biomass and the relative glucose in the medium.

Similarly one can conclude that the reduced carbon source leads to a decrease in apparent viscosity. The rheological parameters can be described in terms of the model of power law fluids for pseudoplastic, which show a nonlinear relationship between shear stress and shear rate.

7. Conclusions

Using the obtained relationships between amino acids were prepared culture media in which, L-threonine limits growth and the amino acids L-leucine and L-methionine is not present in excess, reducing the cost of these amino acid supplementation. The initial concentrations of threonine and glucose, affect the growth of the microorganism. The lack of threonine causes a cessation in growth and a subsequent decrease in biomass, while glucose depending on the initial concentration affects the specific growth rate and inhibits the formation of biomass. Thus the overall process yield is closely related to the initial concentrations of threonine and glucose airlift reactors in the oxygen transfer rate can be increased by increasing the air flow. In studies in the present study we observed that, the use of air flows greater than 1 vvm generate a large amount of foam, making it necessary to use defoamers, in decreasing the solubility of oxygen in the fermentative medium. To maintain adequate oxygenation, in combination with the advantages of the airlift bioreactor, resulted in a prolongation of the maintenance phase, whereby the lysine production rate was constant for a period of time greater than that reported in stirred tank bioreactors. In an airlift, using a minor amount of biomass can generate the same or greater amount of product, reducing the initial amount of amino acids, glucose and ammonium sulfate. Growth models, product formation and substrate consumption correctly predict from the start of the production phase of lysine, to the end of phase constant production rate. It can therefore be used for the simulation of the process from the beginning of fermentation, until the end of the maintenance phase.

Author details

Ana María Mendoza Martínez[1] and Eleazar Máximo Escamilla Silva[2*]

*Address all correspondence to: eleazar@iqcelaya.itc.mx

1 Technological Institute of Madero City, Division of Graduate Studies and Research (ITCM), Madero, México

2 Chemical Departments, Technological Institute of Celaya, Celaya, México

References

[1] Chisti Y, Moo-Young, M. On the calculation of shear rate and apparent viscosity in airlift and bubble column bioreactors. Biotechnol Bioeng. 1989;34:1391-2.

[2] Merchuk JC, Ladwa, N., Cameron, A., Bulmer, M. and Pickett, A. Concentric-tube airlift reactors: Effects of geometrical design on performance. AIChE J. 1994;40:1105-17.

[3] Chisti Y. Pneumatically Agitated Bioreactors in Industrial and Environmental Bioprocessing: Hydrodynamics, Hydraulics and Transport Phenomena. American Society of Mechanical Engineers. 1998;51:33-112.

[4] Joshi J.B. RVV, Gharat S.D., Lee S.S.,. Sparged Loop Reactors. Canadian Journal of Chemical Engineering. 1990;68:705-41.

[5] Petersen E.E. MA. Hydrodynamic and Mass Transfer Characteristics of Three-Phase Gas-Lift Bioreactor Systems. Critical Review in Biotechnology. 2001;21:233-94.

[6] Merchuk JC, Ladwa N., Cameron A., Bulmer M., Pickett A.,. Concentric- Tube Airlift Reactors: Effects of Geometric Design on Performance. AIChE Journal. 1994;40:1105-17.

[7] Gravilescu M, Tudose, R.Z. Modelling mixing parameters in concentric-tube airlift bioreactors. Part I. Mixing time.. Bioprocess Eng,. 1999;20:423-8.

[8] Merchuk JC, Ladwa, N., Cameron, A., Bulmer, M., Pickett, A., Berzin, I. Liquid flow and mixing in concentric tube air-lift reactors. Chem Technol Biotech. 1996;66:172-82.

[9] Merchuk JC, Ladwa, N. Cameron, A. Bulmer, A., Pickett, M. and Berzin, I. Liquid flow and mixing in concentric tube air-lift reactors. Chem Technol Biotech. 1996;66:174-82.

[10] Barnea DT, Y. Fluid Mechanics. In: Cheremisinoff e, editor. Encyclopedia of Fluid Mechanics. Gulf, Houston, Tex., 1986. p. 403-91.

[11] Wallis GB, editor. One Dimensional Two-Phase Flow. New York, : McGraw-Hill; 1969.

[12] Wiswanathan K, editor. Flow Patterns in Bubble Columns. Gulf, Houston, Tex., 1969.

[13] Russell AB, Thomas, C.R., Lilly, M.D.. The influence of vessel height and top section size on the hydrodynamics characteristics of air lift fermenters. Biotechnol Bioeng,. 1994;43:69-76.

[14] Weiland P. Influence of draft tube diameter on operation behaviour of airlift loop reactors. Ger Chem Eng. 1984;7:374-85.

[15] Bello RAR, C.W., Moo-Young, M.. Prediction of the Volumetric Mass Transfer Coefficient in Pneumatic Contactors. Chem Eng Sci,. 1985;40:53-8.

[16] Chisti MY, Moo-Young, M. Airlift reactors: Characteristics, applications and design considerations. Chem Eng Commun,. 1987;60:195-242.

[17] Koide K, Horibe, K., Kawabata, H., Ito, S.,. Gas holdup and volumetric liquid-phase mass transfer coefficient in solid-suspended bubble column with draught tube. J Chem Eng Japan,. 1985;18:248-54.

[18] Merchuk JC, Bulmer, M. Ladwa, N.A., Pickett, M. and Cameron, A, editor. Bioreactor Fluid Dynamics London: Elsevier; 1988.

[19] Siegel MH, Robinson, C.W.,. Applications of airlift gas-liquid solid reactors in biotechnology. Chem Eng Sci 1992;47:3215-29.

[20] Halard B, Kawase, Y., Moo-Young, M.,. Mass transfer in a pilot plant scale airlift column with non-Newtonian fluids.. Ind Eng Chem Res,. 1989;28:243-5.

[21] Mc Manamey WJ, Wase, D.A.J.,. Relationship between the volumetric mass transfer coefficient and gas holdup in airlift fermentors.. Biotechnol Bioeng 1986;28:1446-8.

[22] Merchuk JC, Yunger, R.. The role of the gas—liquid separator of airlift reactors in the mixing process. Chem Eng Sci. 1990;45:2973-5.

[23] Escamilla S, E.M., Dendooven, L., Magaña, I.P., Parra, Saldivar, R., De la Torre, M.. Optimization of Gibberellic acid production by immobilized Gibberella fujikuroi mycelium in fluidized bioreactors. J Biotechnol,. 2000;76:147-55.

[24] Jones A, Pharis, R.P.,. Production of gibberellins and Bikaverin by cells of Gibberella fujikuroi immobilized in carrageenan.. J Ferment Technol,. 1987;65:717-22.

[25] Bailey JE, Ollis, D. F., editor. Transport phenomena in bioprocess systems, Design and analysis of biological reactors. Biochemical engineering fundamentals.. New York: Mc Graw-Hill, ; 1986.

[26] Escamilla S, E.S., Poggi, Varaldo, H., De la torre, Martínez, M., Sánchez, Cornejo, G., Dendooven, L. Selective production of Bikaverin in a fluidized biorreactor with immobilized Gibberella fujikuroi. World J of Microbiol Biotechnol 2001;17:469-74.

[27] Chisti MY, editor. Airlift bioreactor. London-New York: Elsevier Appl. Science, ; 1989.

[28] Quintero RR, editor. Ingeniería bioquímica, Teoría y aplicaciones.. México: Ed. Alambra. ; 1981.

[29] Brito-De la Fuente E, Nava, J.A., López, L.M., Medina, L, Ascanio, G., Tanguy, P.A. Process viscometry of complex fluids and suspensions with helical ribbon agitators. Can J Chem Eng,. 1998;76:689-95.

[30] Shah YT, Kelkar, B.G., Godbole, S.P., Deckwer, W.D.. Design parameters estimations for bubble column reactors.. AIChE J 1982;28:353-79.

[31] Godbole SP, Schumpe, A., Shah, T., Carr, N.L. Hydrodynamics and mass transfer in non-Newtonian solutions in a bubble column.. AIChE J,. 1984;30:213-20.

[32] Gavrilescu M. TRZ. Effects of Geometry on Gas Holdup. Bioprocess Engineering,. 1998;19:37-44.

[33] Abashar M.E. NU, Rouillard A.E., Judd R. Hydrodynamic flow regimes, gas holdup, and liquid circulation in airlift reactors. Ind Eng Chem Res,. 1998;37:1251-9.

[34] Chisti Y, Moo-Young, M. On the calculation of shear rate and apparent viscosity in airlift and bubble column bioreactors. Biotechnol Bioeng,. 1989;34:1391–2.

[35] Freitas C, Teixeira J.A. Hydrodynamic studies in an airlift reactor with an enlarged degassing zone. Bioprocess Engineering. 1998;18:267-79.

[36] Kawase Y. Liquid circulation in external-loop airlift bioreactors.. Biotechnol Bioeng 1989;35:540-6.

[37] Choi KH, Chisti, Y, Moo,Young, M.,. Comparative evaluation of hydrodynamic and gas–liquid mass transfer characteristics in bubble column and airlift slurry reactors. Biochem Eng J,. 1996;62:223-9.

[38] Tobajas M, García, Calvo, E.,. Comparison of experimental methods for determination of the volumetric mass transfer coefficient in fermentation processes. Heat Mass Transfer 2000;36:201-7.

[39] Barboza M, Zaiat M, Hokka, C.O.,. General relationship for volumetric oxygen transfer coefficient (kLa) prediction in tower bioreactors utilizing immobilized cells. Bioprocess Eng,. 2000;22:181-4.

[40] Schügerl KL, J. Oels, U. Bubble column bioreactors. Adv Biochem Eng. 1977;7:1-81.

[41] Akita K Y, F. Gas holdup and volumetric mass transfer coefficient in bubble columns. Effects of liquid properties. Ind Eng Chem Process Des Develop,. 1973;12:76-80.

[42] Prokop A, Janı'k, P., Sobotka, M., Krumphanzi, V.. Hydrodynamics, mass transfer, and yeast culture performance of a column bioreactor with ejector. Biotechnol Bioeng 1983;25:1140-60.

[43] Moo-Young M, Halard, B., Allen, D.G., Burrell, R., Kawase, Y.. Oxygen transfer to mycelial fermentation broths in an airlift fermentor.. Biotechnol Bioeng. 1987;30:746-53.

[44] Al-Masry W.A. DAR. Hydrodynamics and mass transfer studies in a pilot-plant air-lift reactor: non-Newtonian systems. Ind Eng Chem Res,. 1998;37:41-8.

[45] Metz B, Kossen, N.W.F., van Suijdam, J.C.,. The rheology of mould suspensions.. Adv Biochem Eng. 1979;11:103-56.

[46] (INEGI) INdEyG. Demanda de l-Lisina en el estado de Guanajuato. In: Estadistica, editor. Guanajuto2007.

[47] W. L, editor. Amino acids – technical production and use.1996.

[48] Toennies G. Role of aminoacids in postexponential growth.. J Bacteriol 1965;90:438-42.

[49] Van't Riet K. Review of measureing methods and results in nonviscous gas-liquid mass tranfer in stirred vessels. Ind Eng Chem Process Des Develop,. 1979;18:357-64.

[50] Keen RE, Spain, J.D., editor. Computer simulation in biology. Liss. USA: Wiley; 1992.

[51] Singh V. On-line measurements of oxygen uptake in cell culture usig the dynamic method.. Biotechnol Bioeng,. 1996;52:443-8.

[52] Ensari S, Lim, C. H.. Apparent effects of operational variables on the continuous culture of Corynebacterium lactofermentum.. Process Biochemistry 2003;38:1531-8.

[53] Pirt SJ, editor. Principles of microbe and cell cultivation.. London: Blackwell; 1975.

[54] Kiss RD. Metabolic activity control of the L-lisine fermentation by restrained growt fed-batch strategies. Cambridge, MA: M.I.T; 1991.

Chapter 6

Some Remarks on Modelling of Mass Transfer Kinetics During Rehydration of Dried Fruits and Vegetables

Krzysztof Górnicki, Agnieszka Kaleta,
Radosław Winiczenko, Aneta Chojnacka and
Monika Janaszek

Additional information is available at the end of the chapter

1. Introduction

Dehydration operations are important steps in the food processing industry. The basic objective in drying food products is the removal of water in the solids up to a certain level, at which microbial spoilage is minimized. The wide variety of dehydrated foods, which today are available to the consumer (dried fruits, dry mixes and soups, etc.) and the interesting concern for meeting quality specifications, emphasize the need for a thorough understanding of the operation [1].

Dehydrated products can be used in many processed or ready-to-eat foods in place of fresh foods due to several advantages such as convenience in transportation, storage, preparation and use. Dehydrated products need to be rehydrated before consumption or further processing [2]. Rehydration is a process of moistening dry material [3]. Rehydration is usually carried out by soaking the dry material in large amounts of water, although, instead of this, some authors have used air with high relative humidity, either statically or in a drying chamber with air circulation [4].

Three main processes take place simultaneously during rehydration: the imbibition of water into the dried material and the swelling and the leaching of solubles [5]. It is a very complex phenomenon that involves different physical mechanisms such as water imbibition, internal diffusion, convection at the surface and within large open pores, and relaxation of the solid matrix. Capillary imbibition is very important during the early stages, leading to an almost instantaneous uptake of water. Tension effects between the liquid and the solid matrix may also be relevant [6]. In the rehydration process, two main crosscurrent mass fluxes are involved, a water flux from the rehydrating solution to the product, and a flux of solutes (sugars, acids, minerals, vitamins) from the food product to the solution, and the kinetics depends on the immersion medium [2,6,7].

Rehydration is influenced by several factors, grouped as intrinsic factors (product chemical composition, pre-drying treatment, drying techniques and conditions, post-drying procedure, etc.) and extrinsic factors (composition of immersion media, temperature, hydrodynamic conditions) [8]. Some of these factors induce changes in the structure and composition of the plant tissue, which results in the impairment of the reconstitution properties [9]. Therefore equilibrium moisture content at saturation does not reach the moisture content of the raw materials prior to dehydration, indicating that the dehydration procedure is irreversible [1]. Physical and chemical changes that take place during drying affect the quality of the dehydrated product, and by a simple addition of water, the properties of the raw material cannot be restored [10]. Rehydration cannot be simply treated as the reverse process to dehydration [11]. Hence, rehydration can be considered as a measure of the injuries to the material caused by drying and treatments preceding dehydration [12].

Rehydration characteristics are therefore employed as a parameter to determine quality [5]. Optimal reconstitution can be achieved by controlling the drying process and adjustment of the rehydration conditions [13]. The knowledge of the rehydration kinetics of dried products is important to optimise processes from a quality viewpoint since rehydration is a key quality aspect for those dried products that have to be reconstituted before their consumption [14]. The most important aspect of rehydration technology is the mathematical modelling of the rehydration process. Its purpose is to allow design engineers to choose the most suitable operating conditions. The principle of modelling is based on having a set of mathematical equations that can adequately describe the system. The solution of these equations must allow prediction of the process parameters as a function of time. Therefore the use of a simulation model is a valuable tool for prediction of performance of rehydration systems [15].

Many models have been investigated to predict migration of water in foods and, for example, relate moisture content to time. The models are theoretical, empirical, semi-empirical, exponential, and non-exponential ones, and despite the widespread use of computers and their associated softwares, empirical equations are still widely used in view of their simplicity and ease of computations [16]. Theoretical models, however, are based on the general theory of mass and heat transfer laws. They take into account fundamentals of the rehydration process and their parameters have physical meaning. Therefore, theoretical models can give an explanation of the phenomena occurring during rehydration. On the other hand however these models are more difficult in application compared to other mentioned models [15].

The aim of the present chapter was to discuss the suitability of rehydration models and indices for describing mass transfer kinetics during rehydration of dried fruits and vegetables.

2. Mathematical description of rehydration process

2.1. Rehydration indices

There is a large number of research reports in which authors measure the ability of dry material to rehydrate. Results of experiments are expressed in variable ways and quite often the same index is differently named. The most common index used to express rehydration of dry plant tissue is rehydration ratio defined as follows:

$$\text{Rehydration ratio (RR)} = \frac{\text{mass after rehydration}}{\text{mass before rehydration}} \tag{1}$$

Rehydration ratio was used to express the rehydration of the dried products such as carrots [17], mushrooms [18], pears [19], potatoes [20], and coriander leaves [21]. RR is sometimes named rehydration capacity [22,23]. To facilitate a mathematical description of the rehydration phenomenon, the experimental reconstitution data were correlated with time according to a second order polynomial relation [20,24]:

$$RR = a t^2 + b t + c \tag{2}$$

During rehydration together with water acquisition, soluble compounds can be leached. Observed increase in mass is a net result of those processes, and practically gives no information about the amount of absorbed water or the mass of lost solubles. Lewicki in [12] proposed three indices to estimate the rehydration characteristics of dried food. These are the water absorption capacity (WAC), the dry matter holding capacity (DHC), and the rehydration ability (RA).

The water absorption capacity gives information about the ability of the material to absorb water with respect to the water loss during dehydration and varies in the range $0 \leq WAC \leq 1$. The more the water absorption capacity is lost during dehydration the smaller the index. Water absorption capacity is defined by:

$$WAC = \frac{\text{mass of water absorbed during rehydration}}{\text{mass of water removed during drying}} \tag{3}$$

and can be calculated from the formula:

$$WAC = \frac{m_r(1-s_r) - m_d(1-s_d)}{m_o(1-s_o) - m_d(1-s_d)} \tag{4}$$

The dry matter holding capacity is a measurement of the ability of the material to retain soluble solids after rehydration and provides information on the extent of tissue damage and its permeability to solutes. The more the tissue is damaged the smaller the index. The dry matter holding capacity varies in the range $0 \leq DHC \leq 1$. The index is calculated by:

$$DHC = \frac{m_r s_r}{m_d s_d} \qquad (5)$$

The rehydration ability measures the ability of the dried product to rehydrate and shows the total damage of the tissue by drying and soaking processes. The index varies in the range $0 \leq RA \leq 1$. The more the tissue is damaged the smaller is the index. The rehydration ability is given by:

$$RA = WAC \cdot DHC \qquad (6)$$

The indices proposed by Lewicki in [12] were employed to express the rehydration of the dried products such as apples [25,26], *Boletus edulis* mushrooms [4], chestnuts [5], and *Morchella esculenta* (morel) [14].

2.2. Mathematical models

The analysis of the rehydration kinetics can be very useful for optimizing process condition. Many theoretical and empirical approaches have been employed and in some cases empirical models were preferred because of their relative ease of use.

2.2.1. Empirical and semi-empirical models

Among the empirical models, the one proposed by Peleg in [27] is a two parameter, non-exponential equation to describe water transport from the surface to the interior of the solids. The model proposed by Peleg in [27] is as follows:

$$M = M_0 \pm \frac{t}{A_1 + A_2 t} \qquad (7)$$

where A_1 is the Peleg rate constant (s) and A_2 is the Peleg capacity constant.

In Eq. (7) "$\pm$" becomes "+" if the process is absorption or adsorption and "–" if the process is drying or desorption.

The rate of sorption can be obtained from the first derivative of the Peleg equation:

$$\frac{dM}{dt} = \pm \frac{A_1}{(A_1 + A_2 t)^2} \qquad (8)$$

and at the very beginning ($t=0$):

$$\left.\frac{dM}{dt}\right|_{t=0} = \pm \frac{1}{A_1} \tag{9}$$

If time of the process is long enough ($t \to \infty$), the equilibrium moisture content can be calculated by:

$$M_e = M_0 \pm \frac{1}{A_2} \tag{10}$$

Linearization of Eq. (7) gives:

$$\frac{t}{M - M_0} = \pm(A_1 + A_2\, t) \tag{11}$$

allowing for the determination of A_1 and A_2 values by linear regression of experimental data.

Some of the authors correlated A_1 value by means of exponential equation according to an Arrhenius type relationship [4,5]:

$$A_1 = A_0 \exp(-\frac{E_a}{RT}) \tag{12}$$

where A_0 is the constant.

The Peleg [27] model has been widely used due to its simplicity, and has been reported to adequately describe the rehydration of various dried products such as apples [28], bambara [29], candied mango [7], carrots [30], chickpea [31], red kidney beans [32], and wheat [33]. Bilbao-Sáinz et al. in [34] applied Peleg model to volume recovery data assuming the following form of equation:

$$V = V_0 + \frac{t}{A_1 + A_2\, t} \tag{13}$$

Marqes et al. in [35] modified the Peleg model obtaining the following form of equation:

$$\frac{m(t)}{m_d} = \left[\left(\frac{m}{m_d}\right)_{t \to \infty} - \frac{1}{A_2}\right] + \frac{t}{A_1 + A_2\, t} \tag{14}$$

and applied it for modelling of dried tropical fruits rehydration.

Pilosof et al. in [36] proposed empirical, two parameter, non-exponential equation to describe kinetics of water uptake to food powders. The Pilosof-Boquet-Batholomai model [36] is as follows:

$$M = M_0 \pm \frac{A_3 t}{A_4 + t} \tag{15}$$

where A_3 and A_4 are constants.

If time of the process is long enough ($t \rightarrow \infty$), the equilibrium moisture content can be calculated by:

$$M_e = M_0 + A_3 \tag{16}$$

Linearization of Eq. (15) gives:

$$\frac{t}{M - M_0} = \frac{1}{A_3} t + \frac{A_4}{A_3} \tag{17}$$

It can be deduced from Eq. (11) (for absorption or adsorption) and (17) that $A_1 = A_4/A_3$ and $A_2 = 1/A_3$.

The Pilosof-Boquet-Batholomai model [36] has been used by Sopade et al. in [16] for describing water absorption of wheat starch, whey protein concentrate, and whey protein isolate.

Singh and Kulshrestha in [37] proposed empirical, two parameter, non-exponential equation to describe kinetics of water sorption by soybean and pigeonpie grains. The model developed by Singh and Kulshrestha [37] is as follows:

$$M = M_0 \pm \frac{A_5 A_6 t}{A_6 t + 1} \tag{18}$$

where A_5 and A_6 are constants.

If time of the process is long enough ($t \rightarrow \infty$), the equilibrium moisture content can be calculated by:

$$M_e = M_0 + A_5 \tag{19}$$

Linearization of Eq. (18) gives:

$$\frac{t}{M - M_0} = \frac{1}{A_5}t + \frac{1}{A_5 A_6}$$ (20)

It can be deducted from Eq. (11) (for absorption or adsorption), (17) and (20) that $A_1=A_4/A_3=1/(A_5A_6)$, $A_2=1/A_3=1/A_5$ and therefore $A_3=A_5$ and $A_4=1/A_6$.

The Singh-Kulshrestha [37] model has been used in [16] for describing water absorption of wheat starch, whey protein concentrate, and whey protein isolate.

Wesołowski in [38] developed the following empirical, three parameter, exponential equation to describe rehydration of apples:

$$\frac{m(t)}{m_d} = A(B - e^{-Ct})$$ (21)

where A, B, and C are constants.

For a long enough time, equilibrium value is given by:

$$\left(\frac{m(t)}{m_d}\right)_{t\to\infty} = A \cdot B$$ (22)

Equation (21) was also verified when mass has been replaced with moisture content [38,39].

The model proposed by Witrowa-Rajchert in [40] is as follows:

$$\frac{m(t)}{m_d} = A + B\left(1 - \frac{1}{1 + BCt}\right)$$ (23)

where A, B, and C are constants.

It is an empirical, three parameter, non-exponential model. For a long enough time, equilibrium value is given by:

$$\left(\frac{m(t)}{m_d}\right)_{t\to\infty} = A + B$$ (24)

Equation (23) was also verified for the increase of moisture content and volume. Discussed model has been applied for describing the rehydration of apples, carrots, parsleys, potatoes, and pumpkins [40,41].

The probabilistic Weibull model was described first by Dr. Walodi Weibull to represent the distribution of the breaking strength of materials and later to describe the behaviour of systems or events that have some degree of variability [14]. For drying and rehydration processes a two parameter, exponential equation based on the Weibull model is as follows:

$$\frac{M - M_0}{M_e - M_0} = 1 - \exp\left[-\left(\frac{t}{\alpha}\right)^{\beta}\right]$$

(25)

where α is the scale parameter (s) and β is the dimensionless shape parameter. The scale parameter α is a kinetic coefficient. It defines the rate of the moisture uptake process and represents the time needed to accomplish approximately 63% of the moisture uptake process. Different values of α lead to a very different curves: for instance, the higher its value, the slower the process at short times. The shape parameter is a behaviour index, which depends on the process mechanism [42]. Although the Weibull model is empirical one, it was demonstrated recently that the Weibull distribustion has a solid theoretical basis, stemming from physical principles [6].

The Weibull model was found to yield good results in the description of rehydration of a variety of dried foods such as *Boletus edulis* mushrooms [4], *Morchella esculenta* (morel) [14], oranges [43], and ready-to-eat breakfast cereal [42]. Cunha et al. in [44] correlated the scale parameter α value by means of exponential equation according to an Arrhenius type relationship (Eq. (12)).

Marques et al. in [35] modified the Weibull model obtaining the following form of equation:

$$\frac{m(t)}{m_d} = \left(\frac{m(t)}{m_d}\right)_{t\to\infty} + \left[1 - \left(\frac{m(t)}{m_d}\right)_{t\to\infty}\right]\exp\left[-\left(\frac{t}{d}\right)^{\beta}\right]$$

(26)

and applied it for modelling of dried tropical fruits rehydration.

Marabi et al. in [6] modified the Weibull model obtaining the following form of equation:

$$\frac{M - M_0}{M_e - M_0} = 1 - \exp\left[-\left(\frac{t}{\alpha'}\right)^{\beta}\right]$$

(27)

and

$$\alpha' = \frac{L^2}{D_{calc}}$$

(28)

$$D_{calc} = D_{eff} \cdot R_g \tag{29}$$

where R_g is the constant and is a characteristic of the geometry utilized. Marabi et al. in [6] applied Eq. (27) for modelling rehydration of carrots.

The rate of rehydration can be obtained from the semi-empirical first order kinetic model [45]:

$$\frac{dM}{dt} = -k(M - M_e) \tag{30}$$

At zero time M is equal M_0 the moisture content of the dry material, and Eq. (30) is integrated to give the following expression:

$$\frac{M - M_e}{M_0 - M_e} = \exp(-kt) \tag{31}$$

The Arrhenius equation (Eq. (12)) can be employed to describe the temperature dependence of rehydration rate constant k [45-47].

The first order kinetic model has been reported to adequately describe the rehydration of various dried products such as apples, potatoes, carrots, bananas, pepper, garlic, mushrooms, onion, leeks, peas, corn, pumpkins, and tomatoes [1], chickpeas [46], soybeans [47], and tamarind seeds [45].

Misra and Brooker in [48] developed the following empirical, exponential model

$$\frac{M - M_e}{M_0 - M_e} = \exp(-kt^n) \tag{32}$$

(where n is constant) and applied it for modelling the rewetting of dried corn. Equation (32) has been successfully used by Shatadal et al. in [49] to describe the rewetting of dried canola.

Mizuma et al. in [50] modified the first order kinetic model obtaining two form of equations:

$$\frac{dx}{dt} = k(1 - x)^n \tag{33}$$

and

$$\frac{dx}{dt} = k(1 - x)^n (x + a) \tag{34}$$

$$x = \frac{m(t) - m_d}{m_e - m_d} \tag{35}$$

(where a and n are constants) and applied them for modelling of water absorption rate of rice.

2.2.2. Theoretical models

Studies revealed that rehydration is a multifaceted mass transfer process, and uptake is governed by several mechanisms of liquid imbibition in porous media. Rehydration of dried plant tissues is a very complex phenomenon involving different transport mechanisms, including molecular diffusion, convection, hydraulic flow, and capillary flow. One or more mechanisms may occur simultaneously during water or other medium imbibition into a dry food sample [51].

Theoretical models take into account the process basic physical principles. Physically based modelling requires in depth process understanding. As evaluation of some physical properties and complex process interrelationships are very difficult to quantify, the efficiency of these models is typically limited to approximations [51].

Theoretical models describing water absorption in foods are mostly based on the diffusion of water through a porous medium, therefore they assume that liquid water sorption by plant tissue is a diffusion controlled process. If water transport is assumed to take place by diffusion, then the process of rehydration can be described using Fick's second law:

$$\frac{\partial M}{\partial t} = \nabla(D\nabla M) \tag{36}$$

In order to solve the differential equation (36) the following simplifying assumptions were adopted mostly in the literature:

• the initial moisture content in the solid is uniform (the initial condition):

$$M\big|_{t=0} = M_0 \tag{37}$$

• the water diffusion coefficient is constant,

• moisture gradient at the centre of the solid is zero,

• the sample geometry remains constant during the rehydration process,

• external resistance to heat and mass transfer is negligible, i.e. the sample surface attains saturation (equilibrium) moisture content instantaneously upon immersion in absorption media (the boundary condition of the first kind):

$$M\big|_A = M_e \tag{38}$$

- heat transfer is more rapid than mass transfer, so that the process can be assumed isothermal.

Biological materials before drying are cut into small pieces, mostly slices or cubes. They can also be spherical in shape. Therefore Eq. (36) applied to the description of the rehydration of dried material (with the simplifying assumptions mentioned above) takes the following form:

for an infinite plane (slices):

$$\frac{\partial M}{\partial t} = D\frac{\partial^2 M}{\partial x^2} \tag{39}$$

for a finite cylinder (slices):

$$\frac{\partial M}{\partial t} = D\left(\frac{\partial^2 M}{\partial r^2} + \frac{1}{r}\frac{\partial M}{\partial r} + \frac{\partial^2 M}{\partial z^2}\right)$$
$$\left(t > 0; 0 < r < R_C; \ -h < z < +h\right) \tag{40}$$

for a cube:

$$\frac{\partial M}{\partial t} = D\left(\frac{\partial^2 M}{\partial x^2} + \frac{\partial^2 M}{\partial y^2} + \frac{\partial^2 M}{\partial z^2}\right)$$
$$(t > 0; \ -R_c < x < +R_c; \ -R_c < y < +R_c; \ -R_c < z < +R_c) \tag{41}$$

for a sphere:

$$\frac{\partial M}{\partial t} = D\left(\frac{\partial^2 M}{\partial r^2} + \frac{2}{r}\frac{\partial M}{\partial r}\right)$$
$$(t > 0; \ 0 < r < +R_c) \tag{42}$$

(t>0; 0<r<+R$_c$) The initial conditions (Eq. (37)) are following:

for an infinite plane

$$M(x,0) = M_0 = \text{const} \tag{43}$$

for a finite cylinder

$$M(r,z,0) = M_0 = \text{const} \tag{44}$$

for a cube

$$M(x,y,z,0) = M_0 = \text{const} \tag{45}$$

for a sphere

$$M(r,0) = M_0 = \text{const} \tag{46}$$

The boundary conditions of the first kind (Eq. (38)) take the following form:

for an infinite plane

$$M(\pm R_c,t) = M_e = \text{const} \tag{47}$$

for a finite cylinder

$$M(R_c,z,t) = M_e = \text{const} \tag{48}$$

$$\frac{\partial M(0,z,t)}{\partial r} = 0, \; M(0,z,t) \neq \infty \tag{49}$$

$$M(r,h,t) = M_e = \text{const} \tag{50}$$

$$\frac{\partial M(r,0,t)}{\partial z} = 0 \tag{51}$$

for a cube

$$M(\pm R_c, y, z, t) = M_e = \text{const} \tag{52}$$

$$M(x, \pm R_c, z, t) = M_e = \text{const} \tag{53}$$

$$M(x, y, \pm R_c, t) = M_e = \text{const} \tag{54}$$

for a sphere

$$M\left(R_c,\ t\right) = M_e = \text{const} \tag{55}$$

$$\frac{\partial M(0,t)}{\partial r} = 0 \tag{56}$$

An analytical solution of: (i) Eq. (39) at the initial and boundary conditions given by Eqs. (43) and (47), (ii) Eq. (40) at the initial and boundary conditions given by Eqs. (44) and (48)-(51), (iii) Eq. (41) at the initial and boundary conditions given by Eqs. (45) and (52)-(54), and (iv) Eq. (42) at the initial and boundary conditions given by Eqs. (46) and (55)-(56) with respect to mean moisture content as a function of time, take the following form [52]:

for an infinite plane

$$\frac{M(t) - M_e}{M_0 - M_e} = \frac{8}{\pi^2} \sum_{n=0}^{\infty} \frac{1}{(2n+1)^2} \exp\left[-\frac{\pi^2 (2n+1)^2}{4} \cdot \frac{Dt}{R_c^2}\right] \tag{57}$$

for a finite cylinder

$$\frac{M(t) - M_e}{M_0 - M_e} = \frac{32}{\pi^2} \sum_{n=1}^{\infty} \frac{1}{\mu_n^2} \exp\left(-\frac{\mu_n^2 Dt}{R_c^2}\right) \sum_{m=0}^{\infty} \frac{1}{(2m+1)^2} \exp\left[-\frac{\pi^2 (2m+1)^2}{4} \cdot \frac{Dt}{h^2}\right] \tag{58}$$

where μ_n are the roots of the Bessel equation of the first kind of zero order

$$J_0\left(\mu_n\right) = 0 \tag{59}$$

for a cube

$$\frac{M(t) - M_e}{M_0 - M_e} = \frac{512}{\pi^6} \sum_{n=0}^{\infty} \sum_{m=0}^{\infty} \sum_{p=0}^{\infty} \frac{1}{(2n+1)(2m+1)(2p+1)} \exp\left\{-\frac{\pi^2}{4}\left[(2n+1)^2 + (2m+1)^2 + (2p+1)^2\right]\frac{Dt}{R_c^2}\right\} \tag{60}$$

for a sphere

$$\frac{M(t) - M_e}{M_0 - M_e} = \frac{6}{\pi^2} \sum_{n=1}^{\infty} \frac{1}{n^2} \exp\left(-\frac{n^2 \pi^2 Dt}{R_c^2}\right) \tag{61}$$

In order to take into account the necessary number of terms of the series, thirty terms are routinely used in the calculations [4,14], although Sanjuán et al. [11] stated that taking 6-7 terms can be enough. The moisture diffusion coefficient D, also termed effective diffusivity, is an apparent value that comprises all the factors involved in the process. This coefficient is often assumed to be temperature-dependent according to an Arrhenius type relationship (Eq. (12)) [11,53,54]. Fick's equation was also solved considering that the effective diffusivity is moisture-dependent [4]. In that case, the diffusion model cannot be solved analytically. The finite element method (FEM) was used in order to identify the parameters. The relationship between the moisture diffusion coefficient and the moisture content considered was:

$$D = \exp\left(a + bM\right) \tag{62}$$

where a and b are the parameters.

Equation (57) has been reported to adequately describe the rehydration of slices of various dried products such as *Boletus edulis* mushrooms [4], broccoli stems [11], carrots [2,8], and *Morchella esculenta* (morel) [14]. Bilbao-Sáinz et al. in [34] stated that Fick's equation of diffusion (Eq. (58)) was not suitable to model the sorption data of apples (var. Granny Smith). Equation (61) was found to yield good results in the description of rehydration of dried amaranth grains [53], dried date palm fruits [54], and dried soybeans [47].

The mathematical model of rehydration developed by Górnicki in [55] is based on the general theory of mass and heat transfer laws. The model assumes that mass transfer in plant tissue is a diffusion controlled process. The model allows for determination of temperature distribution and concentration distribution of both dry matter and water in time and space inside rehydrated material. The developed model takes into account changeable boundary conditions and changes of material geometry. Six methods of determination of mass transfer coefficients were proposed. The proposed model has been reported to adequately describe the rehydration of slices and cubes both parsley and apples (var. Idared).

Few researches recently embarked on a new approach, which is motivated by the recognition that rehydration of dry food particulates could not be explained and/or modeled solely by a Fickian mechanism. Mechanisms, such as water imbibition, capillarity and flow in porous media, were suggested and are considered relevant for describing the ingress of water into the dried food particulates [51].

Lee et al. in [56] described the rehydration process of freeze-dried fruits (avocado, kiwi fruit, apple, banana, and potato) based on capillary movement of water in the fruit samples. The movement of water through the dried material was assumed to follow capillary motion as described by the Lucas-Washburn equation. The following assumptions were made: (i) the food structure may be simplified as to consist of multi-individual pores, (ii) one dimensional flow, (iii) steady state flow, (iv) fully developed flow, (v) Newtonian fluid with negligible inertia effects, and the following equation was obtained:

$$\frac{d\,h(t)}{d\,t} = \frac{k_1}{h(t)} - k_2 \tag{63}$$

where

$$k_1 = \frac{r\gamma \cos\Theta}{4\eta} \quad \text{and} \quad k_2 = \frac{r^2 g\rho}{8\eta} \tag{64}$$

Lee et al. in [56] stated that the parameter k_1 (m^2s^{-1}) will be the dominant factor of the initial rate of rehydration while the parameter k_2 (m s^{-1}) will become significant during the final state as rehydration approaches equilibrium.

Other researches started to apply the capillary imbibition theory to model the rehydration of foods. Weerts et al. in [57-59] utilized a capillary flow approach to model the temperature and anisotropy effects during the rehydration of tea leafs. Saguy et al. in [60] studied the kinetics of water uptake of freeze-dried carrots and stated that water imbibition followed the general Lucas-Washburn equation. Utilizing different liquid media highlighted, however, the need for model improvement overcoming several discrepancies mainly related to the utilization of a simple "effective" cylindrical capillary and a constant contact angle.

Consequently, due to the complexity of water transport into porous media, the need for further research necessary for the development of the theory and model for the application of capillary imbibition is emphasized [60]. Saguy et al. in [51] elaborated a list of recommended future studies in this field.

3. Discussion of some results of modelling of mass transfer kinetics during rehydration of dried apple cubes

The authors' own results of research are presented in this chapter.

Ligol variety apples used in this study were acquired in local market. The apples were washed in running tap water, hand peeled and the seeds were removed, and then cut into 10 mm cubes thickness using specially cutting machine. Samples were dried on the same day. The fluidized bed drying was carried out using the laboratory dryer constructed in the Department of Fundamental Engineering, Faculty of Production Engineering, Warsaw University of Life Sciences, Warsaw, Poland. The drying chamber consists of a column, which is a Plexiglas cylinder of 12 cm in diameter and 180 cm in height. Drying conditions were 60°C of temperature and 6 m s^{-1} of air velocity. Prior to placing the sample in the drying chamber, the system was run for about one hour to obtain steady conditions. Once the air temperature and fluidization velocity had stabilized, the sample was put into the fluidized bed dryer and the drying begins. Drying was continued until there was no weight change. Experiments were replicated three times. Dried apple cubes were stored in airtight glass containers after dehydration until they were used in the rehydration experiments.

The dried apple sample was rehydrated by immersion in distilled water at 20°C. The ratio of the volume of apple cubes to that of the medium (water) was maintained at 1:25. An initial amount of 10 g of dried apples was used in each trial. The following measurements were replicated three times under laboratory conditions: (i) dry matter of solid changes of the examined samples during rehydration, (ii) volume changes of the examined samples during rehydration, (iii) mass changes of the examined samples during rehydration. Rehydration times were 10, 20, 30, 50, 60, 90, 120, 180, 240, 300 and 360 min. At these specified intervals, samples were carefully removed, blotted with paper towel to remove superficial water, and weighted. Dry matter of solid was determined according to AOAC standards [61]. The mass of samples during rehydration and dry matter of samples were weighted with the electronic scales WPE-300 (RADWAG, Radom, Poland). Maximum relative error was 0.1%. The volume changes of apple cubes during rehydration were measured by buoyancy method using petroleum benzine. Maximum relative error was 5%.

Plot for the variation in mass, dry matter of solid, and volume with time during rehydration are shown in Fig. 1, 2, and 3, respectively. It can be seen from Fig. 1 and 3 that moisture uptake increases with increasing rehydration time, and the rate is faster in the initial period of rehydration and decreased up to the saturation level. This initial period of high water uptake can be attributed to the capillaries and cavities near the surface filling up rapidly [4,62]. As water absorption proceeds, rehydration rates decline due to increased extraction rates of soluble materials [63]. Similar trends have been reported in the previous studies [24,28,64]. It can be observed from Fig. 2 that solute loss increases with increasing rehydration time, and the rate is faster in the initial period of rehydration and decreased up to the saturation level. The explanation of such a course of variation in dry matter of solid with time can be the following. There is an initial steep decrease in solid content because of a high rate of mass transfer (solid gradient). As the solute concentration equilibrated with the environment, the rate of change of solid dry matter is substantially reduced [16]. Similar findings have been noted in the previous studies [2,8,40,55].

The course of rehydration characteristics of apple cubes was described with the following models: the Peleg model (Eq. (7)) [27], the Pilosof-Boquet-Batholomai model (Eq. (15)) [36], the Singh-Kulshrestha model (Eq. (18)) [37], and Witrowa-Rajchert model (Eq.(23)) [40]. The mentioned models were applied for the description of the increase in mass and volume, and the decrease in dry matter of solid. Mass transfer kinetics during rehydration of apple cubes was also modelled using theoretical model based on Fick's second law (Eq. (60)). The variation of dry matter of solid with time and moisture content was described with this model. The goodness of fit of the tested models to the experimental data was evaluated with the determination coefficient (R^2), the root mean square error (RMSE), and reduced chi-square (χ^2). The higher the R^2 value, and lower the RMSE and χ^2 values, the better is the goodness of fit [15,28]. In this study, the regression analyses were done using the STATISTICA routine.

Coefficients of the chosen empirical models and the results of the statistical analyses are given in Table 1.

As can be seen from the statistical analysis results, generally high determination coefficient R^2 were observed for all considered empirical models. The values of RMSE and χ^2 are comparable for all models, although it can be noticed that Witrowa-Rajchert model [40] gave the lowest values of RMSE and χ^2. It turned out from the statistical analyses that the Witrowa-

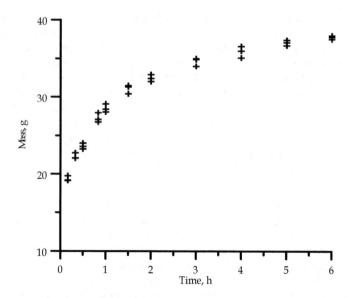

Figure 1. Variation in mass with time during rehydration of apple cubes immersed in distilled water at 20°C

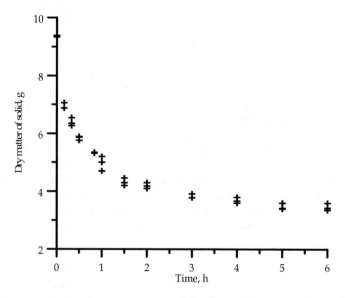

Figure 2. Variation in dry matter of solid with time during rehydration of apple cubes immersed in distilled water at 20°C

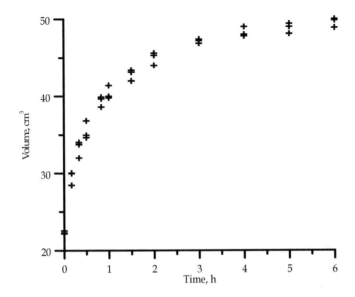

Figure 3. Variation in volume with time during rehydration of apple cubes immersed in distilled water at 20°C

Rajchert model [40] can be considered as the most appropriate. Taking into account values of determination coefficient R^2 it can be, however, stated that all considered models may be assumed to represent the rehydration characteristics. The equilibrium mass and dry matter of solid obtained from the models are in good agreement with the experimental data but only Witrowa-Rajchert model [40] gave appropriate value of equilibrium volume.

Diffusion coefficients estimated from Fick's second law for a cube (Eq. (60)) and the results of the statistical analyses are given in Table 2. Diffusion coefficients are considered constant and because the cubes dimensions changed during the rehydration, four kinds of variables were identified: D/R_c^2, D_1 for R_c=10 mm, D_2 for mean dimension of cube according to time, and D_3 for mean dimension of cube according to moisture content. It can be noticed that diffusion model described the mass transfer kinetics during rehydration of dried apple cubes well. The determined values of D/R_c^2 ($1.37 \cdot 10^{-5}$ s^{-1} and $2.64 \cdot 10^{-5}$ s^{-1}) were found to be lower than the reported in the literature for mushrooms: $4.9 \cdot 10^{-4}$ s^{-1} and $7.9 \cdot 10^{-4}$ s^{-1} [4,14]. The determined values of mass diffusion coefficients was found to be between $2.78 \cdot 10^{-10}$ m^2s^{-1} and $6.60 \cdot 10^{-10}$ m^2s^{-1}. These values are within the general range of 10^{-12}-10^{-8} m^2s^{-1} (mostly about 10^{-10} m^2s^{-1}) for food materials [65,66].

The swelling of dried apple cubes during rehydration was also described using following formula:

$$\frac{V}{V_r} = \left(\frac{M}{M_r}\right)^n \qquad (65)$$

Variable	Model name	Eq. no.	Model parameters	Equilibrium value	R^2	RMSE	χ^2
Mass	Peleg [27]	7	A_1=0.017407 A_2=0.034360	39.104	0.986810	0.970836	1.001431
	Pilosof-Boquet-Batholomai [36]	15	A_3=29.10352 A_4=0.506620	39.407	0.986810	0.986810	0.970836
	Singh-Kulshrestha [37]	18	A_5=29.10352 A_6=1.973866	39.104	0.986810	0.986810	0.970836
	Witrowa-Rajchert [40]	23	A=1.102427 B=2.838299 C=0.622424	39.407	0.987488	0.987488	0.867144
Dry matter of solid	Peleg [27]	7	A_1=-0.04470 A_2=-0.14997	3.332	0.985213	0.204884	0.044601
	Pilosof-Boquet-Batholomai [36]	15	A_3=-6.66822 A_4=0.298063	3.332	0.985213	0.204884	0.044601
	Singh-Kulshrestha [37]	18	A_5=-6.66679 A_6=3.358309	3.333	0.985177	0.204886	0.044602
	Witrowa-Rajchert [40]	23	A=0.932323 B=-0.614312 C=-4.268212	3.154	0.989715	0.168853	0.031271
Volume	Peleg [27]	7	A_1=0.020528 A_2=0.033683	39.689	0.992019	0.762512	0.617763
	Pilosof-Boquet-Batholomai [36]	15	A_3= 29.71020532 A_4= 0.611001699	39.710	0.992033	0.762450	0.617663
	Singh-Kulshrestha [37]	18	A_5=29.71820 A_6=1.634618	39.718	0.992039	0.762449	0.617660
	Witrowa-Rajchert [40]	23	A=1.018117 B=1.317555 C=1.191287	52.156	0.992155	0.740390	0.601227

Table 1. Coefficients of the chosen empirical models and the results of the statistical analyses

Variable	D/R_c^2	D_1	D_2	D_3	R^2	RMSE	χ^2
Dry matter of solid	$2.6407 \cdot 10^{-5}$	$6.6017 \cdot 10^{-10}$	$6.1702 \cdot 10^{-10}$	$5.3756 \cdot 10^{-10}$	0.985583	0.034884	0.000996
Moisture content	$1.3661 \cdot 10^{-5}$	$3.4153 \cdot 10^{-10}$	$3.1921 \cdot 10^{-10}$	$2.7811 \cdot 10^{-10}$	0.99637	0.057613	0.003265

Table 2. Diffusion coefficients ($m^2 s^{-1}$) estimated from the Fick's second law for a cube (Eq. (60)) and the results of the statistical analyses

Such a formula has been used in the literature for modelling drying shrinkage [67]. The model showed a very good fit to the experimental swelling data with a high value of the determination coefficient R^2=0.98920 and low values of root mean square error RMSE=0.01780 and reduced chi-square χ^2=0.00032. The estimated value of swelling coefficient n=0.293.

The coefficients shown in Table 1 were calculated by fitting experimental data to four chosen empirical models. As it was stated in chapter 2.2.1, the coefficients of the Peleg model, the Pilosof-Boquet-Batholomai model, and the Singh-Kulshrestha model are connected between themselves. Table 3 shows the model parameters estimated using these interdependences.

It turned out from the comparison of the results of calculations presented in Tables 1 and 3 that the values of coefficients determined using both methods are almost the same. It can be stated therefore that there is a similarity of these three considered models and their predictive abilities for rehydration of dried apple cubes are identical. The same results have been obtained by Sopade et al. in [16] for describing water absorption of wheat starch, whey protein concentrate, and whey protein isolate. Shittu et al. in [68] observed, however, differences in discussed models predictive ability for the hydration of African breadfruit seeds.

Variable	Model name	Equation no.	Coefficients estimated using coefficients from the following model	Model parameters
Mass	Peleg [27]	7	Pilosof-Boquet-Batholomai [36]	A_1=0.017407518 A_2=0.034360111
	Peleg [27]	7	Singh-Kulshrestha [37]	A_1=0.017407518 A_2=0.034360111
	Pilosof-Boquet-Batholomai [36]	15	Singh-Kulshrestha [37]	A_3=29.10351515 A_4=0.506619958
Dry matter of solid	Peleg [27]	7	Pilosof-Boquet-Batholomai [36]	A_1=-0.044699038 A_2=-0.149964985
	Peleg [27]	7	Singh-Kulshrestha [37]	A_1=-0.044664539 A_2=-0.149997317
	Pilosof-Boquet-Batholomai [36]	15	Singh-Kulshrestha [37]	A_3=-6.666785913 A_4=0.29776892
Volume	Peleg [27]	7	Pilosof-Boquet-Batholomai [36]	A_1=0.020565381 A_2=0.033658468
	Peleg [27]	7	Singh-Kulshrestha [37]	A_1=0.020585496 A_2=0.033649414
	Pilosof-Boquet-Batholomai [36]	15	Singh-Kulshrestha [37]	A_3=29.71819917 A_4=0.611763873

Table 3. Coefficients of the Peleg model [27], the Pilosof-Boquet-Batholomai model [36], and the Singh-Kulshrestha model [37] estimated using the interdependences between the coefficients

4. Conclusions

Four empirical and one theoretical models were investigated for their suitability to describe the mass transfer kinetics during rehydration of dried apple cubes. The determination coefficient, root mean square error, and reduced chi-square method were estimated for all models considered to compare their goodness of fit the experimental rehydration data. All models described the rehydration characteristics of dried apple cubes satisfactorily ($R^2>0.9852$). The empirical Witrowa-Rajchert model [40] and the theoretical model based on Fick's second law can be considered as the most appropriate. Theoretical models give an explanation of the phenomena occurring during rehydration but are difficult in application compared to empirical models. Therefore, if the description of rehydration curves is only needed it is better to apply empirical models. Such need occurs especially in food industry.

The determined values of mass diffusion coefficient was found to be between $2.78 \cdot 10^{-10}$ m²s⁻¹ and $6.60 \cdot 10^{-10}$ m²s⁻¹. These values are within the general range for food materials.

Acknowledgements

The authors are grateful for the financial support from research project No. N N313 780940 from the Polish National Science Centre.

Nomenclature

a, b - constants (Eq.(62))

A, B - constants (Eqs. (21), (22), (23), and (24))

A - surface area (m²)

C - constant (Eqs. (21) and (23))

A_0 - constant (Eqs. (12))

A_1 -constant (Eqs. (7), (8), (9), (11), (12), (13), and (14))

A_2 - constant (Eqs. (7), (8), (10), (11), (13), and (14))

A_3 - constant (Eqs. (15), (16), and (17))

A_4 - constant (Eqs. (15), and (17))

A_5 - constant (Eqs. (18), (19), and (20))

A_6 - constant (Eqs. (18), and (20))

a - constant (Eq. (34))

a, b, c - equation coefficients (Eq. (2))

E_a - activation energy (J mol^{-1})

D - mass diffusion coefficient (effective diffusivity) (m^{2}s^{-1})

D_{calc} - calculated diffusion coefficient (m s^{-1})

D_{eff} - effective diffusion coefficient (m s^{-1})

g - gravitational constant (m s^{-2})

h - half of cylinder heigh, high of liquid rise (m)

k - rehydration rate constant (s^{-1})

k_1 - constant (Eq. (64)) (m^{2}s^{-1})

k_2 - constant (Eq. (64)) (m^{2}s^{-1})

L - characteristic dimension (m)

M - moisture content (dry basis)

M_e - equilibrium moisture content (dry basis)

m - mass kg

n - constant (Eqs. (32), (33), and (34))

n - swelling coefficient (Eq. (65))

r - mean pore radius (m)

r, x, y, z - coordinates (m)

R - universal gas constant (J K^{-1} mol^{-1})

R_c - half of plane or cube thickness, cylinder radius, sphere radius (m)

R_g - constant in Eq.(29)

RMSE - root mean square error

R^2 - coefficient of determination

s - dry matter content (kg d.m. kg^{-1})

T - temperature (K)

t - time (s)

x - water absorption ratio (Eqs. (33), (34), (35))

V - volume (m^3)

α - constant (Eqs. (25) and (26))

α' - constant (Eqs. (27) and (28))

β - constant (Eqs. (25), (26), and (27))

γ - surface tension (N m^{-1})

η - fluid viscosity (Pa s)

θ - advancing liquid constant angle (rad)

ρ - liquid density (kg m^{-3})

χ^2 - reduced chi-square

Subscripts

0 - initial

A - outer surface of body

d - dried

o - before drying

r - rehydrated

Author details

Krzysztof Górnicki, Agnieszka Kaleta, Radosław Winiczenko, Aneta Chojnacka and Monika Janaszek

Faculty of Production Engineering, Warsaw University of Life Sciences, Poland

References

[1] Krokida MK, Marinos-Kouris D. Rehydration Kinetics of Dehydrated Products. Journal of Food Engineering 2003;57(1) 1-7, ISSN 0260-8774.

[2] Nayak CA, Suguna K, Rastogi NK. Combined Effect of Gamma-Irradiation and Osmotic Treatment on Mass Transfer During Rehydration of Carrots. Journal of Food Engineering 2006;74(1) 134-142, ISSN 0260-8774.

[3] Femenia A, Bestard MJ, Sanjuán N, Rosselló C, Mulet A. Effect of Rehydration Temperature on the Cell Wall Components of Broccoli (Brassica oleracea L. Var. italica) Plant Tissues. Journal of Food Engineering 2000;46(3) 157-163, ISSN 0260-87774.

[4] García-Pascual P, Sanjuán N, Bon J, Carreres JE, Mulet A. Rehydration Process of Boletus edulis Mushroom: Characteristics and Modelling. Journal of the Science of Food and Agriculture 2005;85(8) 1397-1404, ISSN 0022-5142.

[5] Moreira R, Chenlo F, Chaguri L, Fernandes C. Water Absorption, Texture, and Color Kinetics of Air-Dried Chestnuts During Rehydration. Journal of Food Engineering 2008;86(4) 584-594, ISSN 0260-8774.

[6] Marabi A, Livings S, Jacobson M, Saguy IS. Normalized Weibull Distribution for Modeling Rehydration of Food Particulates. European Food Research and Technology 2003;217(4) 312-318, ISSN 1438-2385.

[7] Giraldo G, Vázquez R, Martín-Esparza ME, Chiralt A. Rehydration Kinetics and Soluble Solids Lixiviation of Candied Mango Fruit as Affected by Sucrose Concentration. Journal of Food Engineering 2006;77(4) 825-834, ISSN 0260-8774.

[8] Rastogi NK, Nayak CA, Raghavarao KSMS. Influence of Osmotic Pre-Treatments on Rehydration Characteristics of Carrots. Journal of Food Engineering 2004;65(2) pp. 287-292, ISSN 0260-8774.

[9] Taiwo KA, Angarsbach A, Knorr D. Rehydration Studies on Pretreated and Osmotically Dehydrated Apple Slices. Journal of Food Science 2002;67(2) 842-847, ISSN 1750-3841.

[10] Krokida MK, Marolis ZB. Structural Properties of Dehydrated Products During Rehydration. International Journal of Food Science and Technology 2001;36(5) 529-538, ISSN 0950-5423.

[11] Sanjuán N, Simal S, Bon J, Mulet A. Modelling of Broccoli Stems Rehydration Process. Journal of Food Engineering 1999;42(1) 27-31, ISSN 0260-8774.

[12] Lewicki PP. Some Remarks on Rehydration of Dried Foods. Journal of Food Engineering 1998;36(1) 81-87, ISSN 0260-8774.

[13] Marabi A, Thieme U, Jacobson M, Saguy IS. Influence of Drying Method and Rehydration Time on Sensory Evaluation of Rehydrated Carrot Particulates. Journal of Food Engineering 2006;72(3) 211-217, ISSN 0260-8774.

[14] Garcia-Pascual P, Sanjuán N, Melis R, Mulet A. Morchella esculenta (morel) Rehydration Process Modelling. Journal of Food Engineering 2006;72(4) 346-353, ISSN 0260-8774.

[15] Kaleta A, Górnicki K. Some Remarks on Evaluation of Drying Models of Red Beet Particles. Energy Conversion and Management 2010;51(12) pp. 2967-2978, ISSN 0196-8904.

[16] Sopade PA, Xun PY, Halley PJ, Hardin M. Equivalence of the Peleg, Pilosof and Singh-Kulshrestha Models for Water Absorption in Food. Journal of Food Engineering 2007;78(2) 730-734, ISSN 0260-8774.

[17] Wang J, Xi YS. Drying Characteristics and Drying Quality of Carrot Using a Two-Stage Microwave Process. Journal of Food Engineering 2005;68(4) 505-511, ISSN 0260-8774.

[18] Giri SK, Prasad S. Drying Kinetics and Rehydration Characteristics of Micrwave-Vacuum and Convective Hot-Air Dried Mushrooms. Journal of Food Engineering 2007;78(2) 512-521, ISSN 0260-8774.

[19] Komes D, Lovrić T, Kovačević-Ganić K. Aroma of Dehydrated Pear Products. LWT 2007;40(9) 1578-1586, ISSN 0023-6438.

[20] McMinn WAM, Magee TRA. Physical Characteristics of Dehydrated Potatoes – Part II. Journal of Food Engineering 1997;33(1-2) 49-55, ISSN 0260-8774.

[21] Kaur P, Kumar A, Arora S, Ghuman BS. Quality of Dried Coriander Leaves as Affected by Pretreatments and Method of Drying. European Food Research and Technology 2006; 223(2) 189-194, ISSN 1438-2385.

[22] Doymaz I. Air-Drying Characteristics of Tomatoes. Journal of Food Engineering 2007;78(4) 1291-1297, ISSN 0260-8774.

[23] Prothon F, Ahrné LM, Funebo T, Kidman S, Langton M, Sjöholm I. Effects of Combined Osmotic and Microwave Dehydration of Apple on Texture, Microstructure and Rehydration Characteristics. Lebensm. Wiss. u.- Technol.2001;34(2) 95-101, ISSN 0023-6438.

[24] Jambrak AR, Mason TJ, Paniwnyk L, Lelas V. Accelerated Drying of Button Mush-rooms, Brussels Sprouts and Cauliflower by Applying Power Ultrasound and Its Rehydration Properties. Journal of Food Engineering 2007;81(1) 88-97, ISSN 0260-8774.

[25] Atarés L, Chiralt A, González-Martínez C. Effect of Solute on Osmotic Dehydration and Rehydration of Vacuum Impregnated Apple Cylinders (cv. Granny Smith). Journal of Food Engineering 2008;89(1) 49-56, ISSN 0260-8774.

[26] Atarés L, Chiralt A, González-Martínez C. Effect of the Impregnated Solute on Air Drying and Rehydration of Apple Slices (cv. Granny Smith). Journal of Food Engi-neering 2009; 91(2) 305-310, ISSN 0260-8774.

[27] Peleg M. An Empirical Model for the Description of Moisture Sorption Curves. Journal of Food Science 1988;53(4) 1216-1219, ISSN 1750-3841.

[28] Deng Y, Zhao Y. Effect of Pulsed Vacuum and Ultrasound Osmopretreatments on Glass Transition Temperature, Texture, Microstructure and Calcium Penetration of Dried Apples (Fuji). LWT- Food Science and Technology 2008;41(9) 1575-1585, ISSN 0023-6438.

[29] Jideani VA, Mpotokwana SM. Modeling of Water Absorption of Bostwana Bambara Varieties Using Peleg's Equation. Journal of Food Engineering 2009;92(2) 182-188, ISSN 0260-8774.

[30] Markowski M, Zielińska M. Kinetics of Water Absorption and Soluble-Solid Loss of Hot-Air Dried Carrots During Rehydration. International Journal of Food Science and Technology 2011;46(6) 1122-1128, ISSN 0950-5423.

[31] Turhan M, Sayar S, Gunasekaran S. Application of Peleg Model to Study Water Absorption in Chickpea During Soaking. Journal of Food Engineering 2002;53(2) 153-159, ISSN 0260-8774.

[32] Abu-Ghannam N, McKenna B. The Application of Peleg's Equation to Model Water Absorption During the Soaking of Red Kidney Beans (Phaseolus Vulgaris L.). Journal of Food Engineering 1997;32(4) 391-401, ISSN 0260-8774.

[33] Maskan M. Effect of Processing on Hydration Kinetics of Three Wheat Products of the Same Variety. Journal of Food Engineering 2002;52(4) 337-341, ISSN 0260-8774.

[34] Bilbao-Sáinz C, Andrés A, Fito P. Hydration Kinetics of Dried Apple as Affected by Drying Conditions. Journal of Food Engineering 2005;68(3) 369-376, ISSN 0260-8774.

[35] Marques LG, Prado MM, Freire JT. Rehydration Characteristics of Freeze-Dried Tropical Fruits. LWT-Food Science and Technology 2009;42 1232-1237, ISSN 0023-6438.

[36] Pilosof AMR, Boquet R, Batholomai GB. Kinetics of Water Uptake to Food Powders. Journal of Food Science 1985;50(1) 278-282, ISSN 1750-3841.

[37] Singh BPN, Kulshrestha SP. Kinetics of Water Sorption by Soybean and Pigeonpea Grains. Journal of Food Science 1987;52(6) 1538-1541, ISSN 1750-3841.

[38] Wesołowski A. Badanie Suszenia Jabłek Promieniami Podczerwonymi (Investigation on Apple Drying Infrared Radiation). PhD thesis. Warsaw University of Life Sciences, Warsaw, Poland (in Polish); 2000.

[39] Kaleta A, Górnicki K, Kościkiewicz A. Wpływ Parametrów suszenia pod Obniżonym Ciśnieniem na Kinetykę Rehydratacji Suszu z Korzenia Pietruszki (Influence of Vacuum Drying Parameters on Kinetics of Rehyration of Dried Parsley Root). Inżynieria Rolnicza 2006;10 3(78), 69-77, ISSN 1429-7264.29-7264

[40] Witrowa-Rajchert D. Rehydracja Jako Wskaźnik Zmian Zachodzących w Tkance Roślinnej w Czasie Suszenia (Rehydration as an Index of Changes Occurring in Plant Tissue During Drying), Fundacja "Rozwój SGGW", ISBN 83-87660-95-7, Warszawa, Poland (in Polish); 1999.

[41] Kaleta A, Górnicki K. Effect of Initial Processing Methods Used in Conventional Drying Process on the Rate of Getting Equilibrium State in Rehydrated Dried Parsley Root. Annals of Warsaw Agricultural University-SGGW, Agriculture (Agricultural Engineering) 2006;49 9-13, ISSN 1898-6730.

[42] Machado MF, Oliveira FAR, Cunha LM. Effect of Milk Fat and Total Solids Concentration on the Kinetics of Moisture Uptake by Ready-to-Eat Breakfast Cereal. International Journal of Food Science and Technology 1999;34(1) 47-57, ISSN 0950-5423.

[43] Ruiz-Diaz GR, Martínez-Monzó J, Chiralt A. Modelling of Dehydration-Rehydration of Orange Slices in Combined Microwave/Air Drying. Innovative Food Science and Emerging Technologies 2003;4(2) 203-209, ISSN 1466-8564.

[44] Cunha LM, Oliveira FAR, Oliveira JC. Optimal Experimental Desing for Estimating the Kinetic Parameters of Process Described by the Weibull Probability Distribution Function. Journal of Food Engineering 1998;37(2) 175-191, ISSN 0260-8774.

[45] Bhattacharya S, Bal S, Mukherjee RK, Bhattacharya S. Kinetics of Tamarind Seed Hydration. Journal of Food Engineering 1997;33(1-2) 129-138, ISSN 0260-8774.

[46] Gowen A, Abu-Ghannam N, Frias J, Oliveira J. Modelling of Water Absorption Process in Chickpeas (Cicer arietinum L.) – The Effect of Blanching Pre-Treatment on Water Intake and Texture Kinetics. Journal of Food Engineering 2007;78(3) 810-819, ISSN 0260-8774.

[47] Gowen A, Abu-Ghannam N, Frias J, Oliveira J. Influence of Pre-Blanching on the Water Absorption Kinetics of Soybeans. Journal of Food Engineering 2007;78(3) 965-971, ISSN 0260-8774.

[48] Misra MK, Brooker DB. Thin-Layer Drying and Rewetting Equations for Shelled Yellow Corn. Transactions of the ASAE 1980;23(5) 1254-1260, ISSN 0001-2351.

[49] Shatadal P, Jayas DS., White NDG. Thin-Layer Rewetting Characteristics of Canola. Transactions of the ASAE 1990;33(3) 871-876, ISSN 0001-2351

[50] Mizuma T, Tomita A, Kitaoka A, Kiyokawa Y, Wakai Y. Water-Absorption Rate Equation of Rice for Brewing Sake. Journal of Bioscience and Bioengineering 2007;103(1) 60-65, ISSN 1389-1723.

[51] Saguy I S, Marabi A, Wallach R. New Approach to Model Rehydration of Dry Food Particulates Utilizing Principles of Liquid Transport in Porous Media. Trends in Food Science and Technology 2005;16(10) 495-506, ISSN 0924-2244.

[52] Crank J. Mathematics of Diffusion, 2nd ed., Clarendon Press, ISBN 0198534116, Oxford, UK; 1975.

[53] Calzetta Resio AN, Aguerre RJ, Suárez C. Study of Some Factors Affecting Water Absorption by Amaranth Grain During Soaking. Journal of Food Engineering 2003;60(4) 391-396, ISSN 0260-8774.

[54] Falade KO, Abbo ES. Air-Drying and Rehydration Characteristics of Date Palm (Phoenix dactylifera L.) Fruits. Journal of Food Engineering 2007;79(2) 724-730, ISSN 0260-8774.

[55] Górnicki K. Modelowanie Procesu Rehydratacji Wybranych Warzyw i Owoców (Modelling of Selected Vegetables and Fruits Rehydration Process), Wydawnictwo SGGW, ISBN 978-83-7583-325-6, Warszawa, Poland (in Polish); 2011.

[56] Lee KT, Farid M, Nguang SK. The Mathematical Modelling of the Rehydration Characteristics of Fruits. Journal of Food Engineering 2006;72(1) 16-23, ISSN 0260-8774.

[57] Weerts AH, Lian G, Martin DR. Modeling Rehydration of Porous Biomaterials: Anisotropy Effects. Journal of Food Science 2003;68(3) 937-942, ISSN 1750-3841.

[58] Weerts AH, Lian G, Martin DR. Modeling the Hydration of Foodstuffs: Temperature Effects. AIChE Journal 2003;49(5) 1334-1339, ISSN 1547-5905.

[59] Weerts AH, Martin DR, Lian G, Melrose JR. Modelling the Hydration of Foodstuffs. Simulation Modelling Practice and Theory 2005;13(2) 119-128, ISSN 1569-190X.

[60] Saguy IS, Marabi A, Wallach R. Liquid Imbibition During Rehydration of Dry Porous Foods. Innovative Food Science and Emerging Technologies 2005;6(1) 37-43, ISSN 1466-8564.

[61] AOAC Official Methods of Analysis. Arlington, VA: Association of Official Analytical Chemists (No. 934.06); 2003.

[62] Cunningham SE, McMinn WAM, Magee TRA, Richardson PS. Effect of Processing Conditions on the Water Absorption and Texture Kinetics of Potato. Journal of Food Engineering 2008;84(2) 214-223, ISSN 0260-8774.

[63] Abu-Ghannam N, McKenna B. Hydration Kinetics of Red Kidney Beans (Phaseolus vulgaris L.). Journal of Food Science 1997;62(3) 520-523, ISSN 1750-3841.

[64] Planinić M, Velić D, Tomas S, Bilić M, Bucić A. Modelling of Drying and Rehydration of Carrots Using Peleg's Model. European Food Research and Technology 2005;221(3-4) 446-451, ISSN 1438-2385.

[65] Doulia D, Tzia K, Gekas V. A Knowledge Base for the Apparent Mass Diffusion Coefficient (DEFF) of Foods. International Journal of Food Properties 2000;3(1) 1-14, ISSN 1532-2386.

[66] Maroulis ZB, Saravacos GD, Panagiotou NM, Krokida MK. Moisture Diffusivity Data Compilation for Foodstuffs: Effect of Material Moisture Content and Temperature. International Journal of Food Properties 2001;4(2) 225-237, ISSN 1532-2386.

[67] Górnicki K, Kaleta A. Modelling Convection Drying of Blanched Parsley Root Slices. Biosystems Engineering 2007;97(1) 51-59, ISSN 1537-5110.

[68] Shittu TA, Awonorin SO, Raji AO. Evaluating Some Empirical Models for Predicting Water Absorption in African Breadfruit (Treculia Africana) Seeds. International Journal of Food Properties 2004;7(3) 585-602, ISSN 1532-2386.

Chapter 7

General Aspects of Aqueous Sorption Process in Fixed Beds

M. A. S. D. Barros, P. A. Arroyo and E. A. Silva

Additional information is available at the end of the chapter

1. Introduction

Adsorption as well as ion exchange share many common features. Although such phenomena are distinct (ion exchange is a stoichiometric process) they can be generically denominated as sorption processes. Then, as already known, sorption is a well-established technology in chemical engineering.

The sorption ability of different sorbents is strongly dependent on the available surface area, polarity, contact time, pH and the degree of hydrophobic nature of the adsorbent and adsorbate (Suzuki, 1990). Therefore, the selection of the sorbent is the first step in such investigation. Studies should be firstly focused on batch systems in equilibrium and isotherms should be constructed.

Equilibrium condition is attained when the concentration of the solute remains constant, as a result of zero net transfer of solute sorbed and desorbed from sorbent surface. The equilibrium sorption isotherms describe these relationships between the equilibrium concentration of the sorbate in the solid and liquid phase at constant temperature. Experimental data may provide different isotherm shapes such as: Linear, Favorable, Strongly favorable, Irreversible and Unfavorable (McCabe et al., 2001) as shown in Figure 1.

The linear isotherm starts from the origin. Although it does not show a selectivity behavior, the sorbent related to this isotherm sometimes is chosen because the linear isotherm facilitates the column modeling, mainly when equilibrium data is added to the phenomenological dynamic model (Helferich, 1995).

Sorbents that provide isotherms with convex upward curvature are denominated as favorable and strongly favorable. These shapes are often selected for dynamic studies because they show the sorbent selectivity to the sorbate of interest. Nevertheless, they are not convenient when regeneration is required because it may be related to an unfavorable isotherm and quite long mass-transfer zones in the bed (McCabe et al., 2001). Moreover, the famous Langmuir model provides good adjustments to strongly favorable isotherms as it forms a plateau that represents in many cases, the monolayer sorption. Here, it must be remembered that the equilibrium equation is used in the modeling of fixed bed.

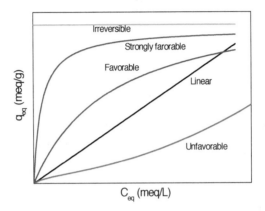

Figure 1. Adsorption isotherms

Finally, it may be emphasized that the Langmuir isotherm is derived assuming a uniform surface, which is many cases is not rigorously valid. This relation works fairly well, even with sorbents with high heterogeneity such as zeolites, clays or activated carbon.Here some aspects should be stressed. The first one is related to the model itself. Sometimes the Langmuir model represents the experimental data although it is known that are different sites involved in the sorption process such as the ones located in the supercages or in the sodalite cages of NaX zeolite (Barros et al., 2004). On the other hand, sorbent may have high heterogeneity and only one type of site is effective in the sorption process. Then, the experimental data may be fitted successfully to such model and the assumptions previously considered are still valid. This may be case of sorption of large molecules such as dyes.

The limiting case of the extremely favorable isotherm is the irreversible sorption represented by a horizontal line, that means a constant amount of sorbed compound. When the irreversible isotherm is obtained no regeneration is possible.

Designing of dynamic sorption processes also takes into account the multicomponent effluent. It means that studies of batch competitive systems should be relevant. Sorbents that have affinity to different sorbates may be less effective in removing the one of interest. It these cases, higher packed beds are required. The multicomponent equilibrium data are obtained considering the initial multicomponent solution. Many models have been proposed.

Some of them are derived from the single Langmuir model. As it happens with the single sorption, even when the ion exchange phenomenon is involved, release of the out-going ion is neglected (Misak 2000; Sprynskyy et al., 2006).

The Binary Langmuir Model assumes a homogeneous surface with respect to the energy of adsorption, no interaction between adsorbed species and that all adsorption sites are equally available to all adsorbed species The Noncompetitive Langmuir Model considers that the concentration of the sorbate of interest in the sorbent depends on the concentration of the respective specie in the fluid phase only. In this case, monocomponent Langmuir can be applied for each one of the species in solution (Sánchez et al., 1999). In both cases, the Binary Langmuir Model and the Noncompetitive Langmuir Model, as there is no competition to the same sites, no overshooting is observed in the breakthrough curve. Overshooting will be discussed later.

The Langmuir Type Model was developed to describe the noncompetitive inhibition during enzymatic kinetic studies. According to the adsorption point of view such model may be applied when a synergism effect is presented due to the existence of sites containing both species (Bailey and Ollis, 1986; Sánchez et al., 1999). It means that in the dynamic sorption more adsorption occurs than the one it was expected, and the uptake on the sorbate of interest is promoted by the presence of other sorbates. No overshooting is expected to occur in this system.

Jain and Snoeyink (1973) have proposed an adsorption model for binary mixtures based on the hypothesis that a part of adsorption occurs without competition. In other words it means that there is at least one type of sorption site where the sorbate of interest is preferentially retained with no competition. In other types of sites, competition occurs. Then, it may be supposed that a slightly overshooting curvature may be evidenced in the breakthrough curve.

Myers and Prausnitz (1965) developed the ideal adsorbed solution theory (IAST) based on the Gibbs adsorption isotherm. The IAST model has the main advantage to predict the equilibrium in multicomponent systems in microporousmaterials through the single isotherms only. However, although the model allows its application in many complex mixtures of solutes, the IAST model limits its use in the range of concentration where the single equilibrium data were obtained. It has been used to predict the sorption process in gas phase, mainly in diluted systems.

A non-predictive thermodynamic approach called RAST (Real Adsorbed Solution Theory) extends the IAST theory to more concentrated solutes using the corresponding activity coefficients.

Finally, when the ion exchange process is much more pronounced then the adsorption process, it may be applied the Ion Exchange Model taking into account the Mass Action Law. In typical ion exchangers such as zeolites this mode are applied with successful results.

After selecting the sorbent and investigating the sorption mechanism, the second step is the dynamic studies in the fixed bed system.

2. Dynamic studies in fixed bed systems

2.1. The breakthrough curve

Most ion-exchange operations, whether in laboratory or in plant-scale processes, are carried out in columns. A solution is passed through a bed of sorbent beads where its composition is changed by sorption. The composition of the effluent and its change with time depend on the properties of the sorbent (as already discussed), the composition of the feed, and the operating conditions (flow rate, temperature etc.). Plots of the ratio C/C_0 (outlet sorbate concentration/sorbate feed concentration) versus time are denominated as breakthrough curves.

As the run stars, most of the mass transfer takes place near the inlet of the bed, where the fluid first contacts the sorbent. If the solid contains no sorbate at the start, the concentration in the fluid drops exponentially to zero before the end of the bed is reached. This concentration profile as well as the breakthrough curve is shown in Figure 1. As the run proceeds, the solid near the inlet is nearly saturated, and most of the mass transfer takes place further from the inlet. The concentration gradient is S shaped. The region where most of the change in concentration occurs is called the mass-transfer zone. This is the real behavior of mass transfer process in fixed beds. When the axial or radial mass transfer resistances are neglected sorption occurs homogeneously and this is the ideal case. In fact, mass transfer resistances can be minimized but not effectively eliminated. Comments about such phenomenon will be better detailed. The limits of the breakthrough curve are often taken as C/C_0 values of 0.05 to 0.95, unless any other recommendation is fixed. They are related to the breakpoint (t_b, C_b) and saturation point (t_s, C_s), respectively. In most of cases $C_s \approx C_0$. This is the case of wastewater treatment of highly toxic sorbates. When the concentration reaches the limiting permissible value, say, 1 ppm, it is considered the break point. The flow is stopped, the column is regenerated and the inlet concentration is redirected to a fresh sorbent bed.

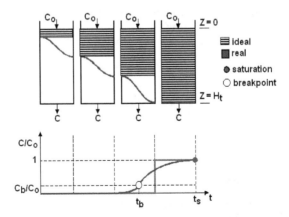

Figure 2. Breakthrough curve for the sorption process in fixed beds Co is the concentration of the inlet solution, C_b is the concentration of the breakthrough, t_b is the breakpoint time and t_s is the saturation time.

2.2. Mass transfer in fixed beds

In fixed-bed ion exchange, the concentration in the fluid phase and in the solid phase changes with time as well as with position in the bed. The transfer process is described by the overall volumetric coefficient ($K_c a$), obtained from a metal material balance in the column assuming irreversible sorption as proposed in McCabe et al. (2001):

$$K_c a = \frac{N \cdot u_o}{H_t} \tag{1}$$

where N is the overall number of transfer units, H_t is the bed length and u_o is the superficial velocity of the fluid.

In fact, Eq. 1 can be used for modeling the breakthrough curves, if the batch isotherms can be considered as irreversible.

The overall number of transfer units may be obtained graphically by plotting C/C_o versus $N(\tau-1)$, where $N(\tau-1) = 1+\ln(C/C_o)$ (McCabe et al., 2001). Parameter τ is dimensionless time defined as

$$\tau = \frac{u_o C_o\left(t - \frac{\varepsilon}{u_o} H_t\right)}{\rho_p (1 - \varepsilon) W_{sat} H_t} \tag{2}$$

The term $\frac{\varepsilon}{u_o} H_t$ in Eq. 2 is the time required to displace fluid from external voids in the bed, which is normally negligible. The product is the total amount of metal fed per unit cross section of the bed up to time t and $\rho_p (1 - \varepsilon) W_{sat} H_t$ is the capacity of the bed, which is equal to the time equivalent to total stoichiometric capacity of the packed-bed tower (t_t).

The time equivalent to usable capacity of the bed (t_u) and the time equivalent to total stoichiometric capacity of the packed-bed tower (t_t) if the entire bed reaches equilibrium are provided by a mass balance in the column and they are easily determined by (Geankoplis 1993):

$$t_u = \int_0^{t_b}\left(1 - \frac{C}{C_o}\right)dt \tag{3}$$

$$t_t = \int_0^{\infty}\left(1 - \frac{C}{C_o}\right)dt \tag{4}$$

where t_b is the breakpoint time.

If time t is assumed to be the time equivalent to the usable capacity of the bed (t_u) up to t_b, parameter τ may be simplified to t_u/t_t. This ratio is the fraction of total bed capacity or length utilized to the breakpoint (Geankoplis, 1993). Hence, the length of unused bed (H_{UNB}) is the unused fraction times the total length (H_t).

$$H_{UNB} = \left(1 - \frac{t_u}{t_t}\right) H_t \qquad (5)$$

H_{UNB} is assumed to be constant and, as a consequence, an important tool when scaling-up processes (McCabe et al., 2001). Unfortunately it is not always true. A constant mass-transfer zone is valid for ideal sorption systems associated with sorbates of small molecular diameter and simple structures (Walker and Weatherly, 1997). Changes in pH speciation through the column may also change the MTZ (Gazola et al., 2006). It probably happens because the rate at which the sorption zone travels through the bed decreases with bed height (Walker and Weatherly, 1997). Therefore, it may be concluded that the hypothesis of a constant length of mass transfer zone for the same feed concentration can be acceptable depending on the variation of the bed height and specific sorbates.

H_{UNB} represents the mass-transfer zone (MTZ). Small values of this parameter mean that the breakthrough curve is close to an ideal step with negligible mass-transfer resistance. Moreover, in the ideal condition, no axial dispersion would occur. The velocity profile would be analogous to the one observed in a Plug Flow Reactor and the ideal breakthrough curve would be the response to a positive-step test, called the Cumulative Distribution Function or F curve. In the ideal breakthrough curve H_{UNB} is zero. This condition is never reached although it is recommended to operate the column as close as possible. The closer the column is operated to the ideal condition, the more efficient is the mass transfer zone. Therefore, the ideal situation means $H_{UNB} = 0$. In experimentally effective situation $0 < H_{UNB} < H_t$. If $H_{UNB} > H_t$ in the very beginning of the run the sorbate is presented in the outlet solution, the sorption process is highly inefficient, mainly if the sorption unit has been used for wastewater treatment. An increase in H_t is recommended.

Besides the minimum MTZ, some other mass transfer parameters may be used to identify, quantitatively, the condition with the lowest resistances over the operational conditions investigated.

One of this parameter is the residence-time distribution (RTD). According to Fogler (2004), the RTD is determined experimentally by injecting a tracer in the column at some time t=0 and then measuring the outlet concentration as a function of time. In this case, the tracer is used to determine RTD in reactors and must be nonreactive, easily detectable, completely soluble to the mixture and more important: it should not adsorb on the walls or any other surface of the reactor. If the injection of a tracer is given as a step injection, the response is the F curve which shape is analogous to the breakthrough curve as seen in Figure 2. Then, it is possible to obtain the average residence time (t_{res}). Correlation between the F curve from the non-ideal concept and the sorption process in fixed beds can be done only in terms of the graphical shape. No more similarities are corrected.

The average residence time (t_{res}) of the fluid in the column is estimated based on principles of probability as follows (Hill, 1977):

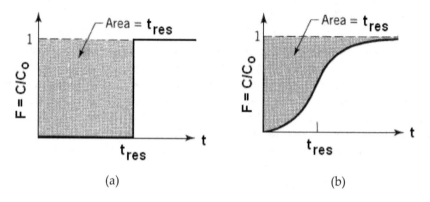

Figure 3. Breakthrough curves considering the F response of a step input: a) ideal situation ($H_{UNB} = 0$) ; b) real situation ($H_{UNB} > 0$).

$$t_{res} = \int_0^\infty t \, dF(t) \tag{6}$$

where F(t) is the weight fraction of the effluent of an age less than t, which is equivalent to C/C_o for breakthrough curves (Barros et al, 2003).

An indirect measure of how far from the optimum operational condition the column operates is expressed by the operational ratio (R) as (Barros et al., 2003)

$$R = \left| \frac{t_{res} - t_u}{t_u} \right| \tag{7}$$

Values of parameter R close to zero indicate that the operational conditions imposed are near the ideal condition, i.e., the optimal region of operation. Therefore, this difference may contribute to the selection of the best operational conditions in the column design.

With the average residence time it is also possible to evaluate the variance in the breakthrough curve (Hill, 1977), which is given by:

$$\sigma^2 = \int_0^\infty t^2 \left(\frac{F(t)}{dt} \right) - t_{res}^2 \tag{8}$$

The dimensionless variance should be calculated as

$$\sigma_\theta^2 = \frac{\sigma^2}{t_{res}^2} \tag{9}$$

Determination of this parameter is useful to estimate the axial dispersion in the packed bed. Values of dimensionless variance close to zero mean that the behavior of the velocity profile in packed beds is close ideal plug-flow with negligible axial dispersion.

Finally, through a mass balance it is possible to obtain the amount of sorbate retained up to the breakpoint time (U^{t_b}) and up to the saturation time (U^{t_s}).

$$U^{t_b}=t_u.f_R.C_o \tag{10}$$

$$U^{t_s}=t_t.f_R.C_o \tag{11}$$

where t_u and t_t are obtained through Eqs. 3 and 4, f_R is the flow rate and C_o is the inlet concentration.

2.3. Minimum mass transfer resistances in the fixed bed

Operate the fixed bed with minimum mass transfer resistances is quite advantageous. It maximizes the sorption process as more sorption sites are available to the dynamic process. Therefore, optimizing the operational conditions is highly recommended. It can be done investigating a range of particle diameters of the sorbent or different flow rates in the column. In some cases, temperature should be also investigated. It is expected to obtain higher amount of uptaken sorbate with increasing temperatures when chemisorption and/or ion exchange are presented as the more significant mechanism. Nevertheless, if the retention is due to physisorption phenomenon, an increase in temperature is disadvantageous.

The optimal operational condition is the one that minimizes, as close as possible, the film and particle resistances, that means to maximize the pore diffusion and the solid diffusion. Figure 3 gives an idea about the mass transfer mechanism in a fixed bed.

Therefore, when the film and particle resistances are minimized it is seen that:

The mass transfer zone, the operational ratio and the dimensionless variance are minimum;

The closer the system is of the optimal condition, the closer the breakpoint time is to the saturation time;

The sorbate uptake up to breakpoint time (U^{t_b}) and up to saturation time (U^{t_s}) are the maximum. It happens when there is at least a favorable isotherm;

In most of cases, a steep breakthrough curve is observed.

Figure 4 shows an example of breakthrough curves at different flow rates. It is seen that at 11 mL/min the breakthrough curve is almost a step curve. Table 1 presents the numerical data and it is total agreement with Figure 4. It means that at this flow rate the minimum mass transfer resistance was reached considering the range investigated.

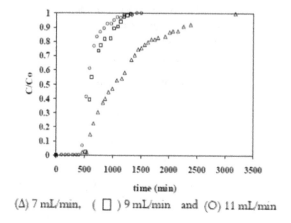

$$(\Delta) \ 7 \ mL/min, \quad (\ \square \ ) \ 9 \ mL/min \quad and \quad (O) \ 11 \ mL/min$$

Figure 4. Breakthrough curves for Cr (III) uptake in NaA zeolite (Barros et al., 2002)

Flow rate (mL/min)	H_{UNB} (cm)	R	$\sigma_\theta^2 = \dfrac{\sigma^2}{t_{res}^2}$	$U_{Cr}^{t_b}\Big\vert_{CEC_{bed}}$
7.0	3.9	1.3	0.3	0.25
9.0	1.7	1.0	0.3	0.30
11.0	1.0	0.4	0.1	0.31

$U_{Cr}^{t_b}\Big\vert_{CEC_{bed}}$ = ratio of amount of Cr (III) retained up to the breakpoint and the cation exchange capacity of the zeolite in the bed.

Other examples can be found in Pereira et al. (2006).

Table 1. Mass-transfer parameters obtained for Cr(III) uptake in NaA zeolite packed bed (Barros et al., 2002)

2.4. Sorption in competing systems

When the feed solution is composed by different sorbates, the competition for the sorption sites may occur. Uptake of the sorbate of interest is well evidenced by the dynamic capacity of the column. Thus, the term $U_{i\text{-}mix}^{t_b}$ is defined as the amount of sorbate uptaken at t_b prior to the sorbate i breakpoint (Valdman et al., 2001).

The effect of competitive systems on the sorption process may be represented by the ratio of the uptake capacity for the sorbent i in multicomponent solution and in single solution, that is the removal ratio $\left(U_{i\text{-}mix}^{t_b} \Big/ U_{i\text{-}single}^{t_b} \right)$ -Mohan and Chander, (2001).

Thus if the removal ratio:

> 1: uptake is promoted by the presence of other sorbates;

=1: no interaction exists between the sorbates;

< 1: uptake is suppressed by the presence of other sorbates.

Therefore, the evaluation of the removal ratio at the time equivalent to usable capacity of the bed t_u for sorbate i may be useful to investigate the influence of different sorbates on the uptake of such sorbate (Barros et al., 2006).

Moreover, in the competing systems, sometimes the breakthrough curves for the competing sorbates may have values higher than one that is called overshooting. When the ion exchange phenomenon is dominant, it is also called sequential ion exchange (Barros et al., 2006). Figure 5 shows an example of a sequential ion exchange. The ion exchange preferentially occurs up to the occupation of all available sites. As the sorption proceeds, the competing sorbates are released at different running times with continued feed of the multicomponent solution. Potassium ions are smaller than the other ions. Then, they diffuse faster into the ion exchanger framework and are firstly retained. Calcium ions are larger and probably face some difficulties in diffusing as fast as the potassium ions. Nevertheless, once they diffuse, they are preferentially retained due to its higher charge, displacing the potassium ions. Finally, chromium ions are the largest ones with the highest charge. They diffuse slowly into the exchanger framework. They displace the calcium ions. The consequence of such process is the presence of two overshooting waves: one related to the release of potassium ions and another related to the release of the calcium ions. In this competing ion exchange process it is seen that $\left(U_{Cr-mix}^{t_b}\middle/U_{Cr-single}^{t_b}\right)$ is 0.79, which means that there is a decrease of 21% in the chromium uptake. Probably the presence of potassium and mainly the presence of calcium ions suppressed the chromium uptake due to competition towards the exchanging sites located in the large cages of the zeolite NaY.

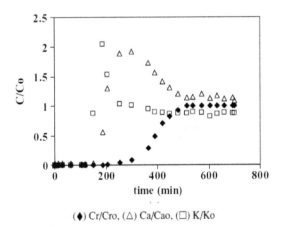

$(\blacklozenge)$ Cr/Cro, $(\triangle)$ Ca/Cao, $(\square)$ K/Ko

Figure 5. Breakthrough curves for competitive system Cr/Ca/K in NaY zeolite (Barros et al.,2006)

Overshooting may occur in different sorbents such as activated carbons (Mohan and Chander, 2001), biosorbents (Sağ et al., 2000) or even ion exchanger membranes (Labanda et al., 2011). It depends on the selectivity of the sorbent, the sorption mechanism (that can be seen through the isotherms) and on the operational conditions imposed. If the operational condition is the one that minimizes the mass transfer resistances of one sorbate is imposed, it is expected a pronounced overshooting when the optimal operational conditions of the other sorbates are different from the one of interest. When the optimal conditions are close, there should be some overshooting depending of the selectivity provided by the isotherms and the concentration of the sorbates, as shown in Figure 6.

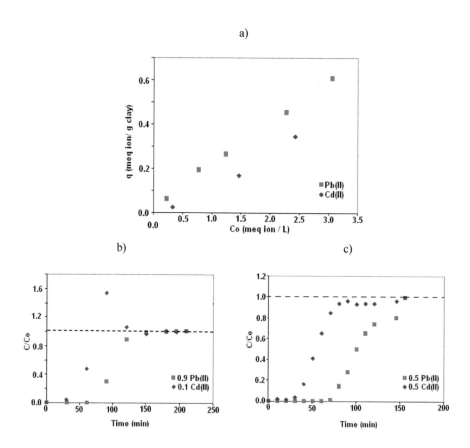

Figure 6. a) Dinamic isotherm (see item 2.5) of Pb (II)/Cd (II) in bentonite clay pretreated with calcium at 30 °C and 2 mL/min; b) breakthrough curve at 2 mL/min and total concentration of 3.0 meq/L, 90% Pb (II), 10% Cd (II); c) breakthrough curve at 2 mL/min and total concentration of 3.0 meq/L, 50% Pb (II), 50% Cd (II).

In Figure 6a it is seen the bicomponent dynamic isotherm of Cd (II) and Pb(II) ions in bentonite clay from Boa Vista-PB-Brazil. The clay sample was pretreated with calcium in order to obtain, as far as possible, a homoionic sample. The concept of dynamic isotherm will be discussed in section 2.5. In this system, the flow rate of 2 mL/min may be considered as the optimal condition that minimize the mass transfer resistances of both Cd(II) and Pb(II) uptake. It is seen that the clay sample has more affinity to lead ions, probably due to its electronegativity (Kang et al., 2004).

Figures 6b and 6c show the breakthrough curves for two different inlet percentages of cadmium and lead and the same total concentration of 3 meq/L. In the first case the inlet concentration was composed by 90% Pb (II), 10% Cd (II) (2.7 meq/L of Pb (II) and 0.3 meq/L of Cd (II). In the second case the composition was 50% for each ion, or 1.5 meq/L of Pb (II) and 1.5 meq/L of Cd (II). The overshooting is clearly observed in Figure 6b due to the higher selectivity to lead ions. When cadmium concentration was increased no overshooting was observed although, according to the breakpoint times, it is also seen the preference of lead as its breakpoint occurs later.

2.5. Dynamic isotherms

Most separation and purification processes that employ the sorption technology use continuous-flow columns. In order to understand and to model the dynamic process, a deep knowledge of the equilibrium in the fixed bed is essential. For the complete comprehension of the whole process, information about the equilibrium and the kinetics of the process should be combined with the mass balance. Kinetic investigations in a column are almost exclusively restricted to those processes of which the equilibria either can be represented by a linear or a Langmuir isotherm or satisfy the law of mass action. However, the results of this study of industrially applicable commercial products have shown that the above equilibria usually fail to fit the experimental results adequately. Furthermore, the customary procedures of determining the equilibria in batch systems are not, in general, applicable to all types of sorbents (Klamer and Van Krevelen, 1958).For example, the batch isotherm was not able to represent correctly, the breakthrough curve of Cr (III) sorption in NaX zeolite modeled by a phenomenological model. Such model will be discussed in section 2.6. Figure 7 presents the batch isotherm of Cr (III) in zeolite NaX and its use in a LDF model. It is clearly seen that it is totally inadequate.

The dynamic isotherm surged to better describe the sorption mechanisms involved in a dynamic process where kinetics and equilibrium acts simultaneously. In fixed beds, solution is fed continuously, and, at equilibrium, concentration and pH are equal to their respective feed values that do not occur in batch isotherms. Its procedure is based on the breakthrough data. Each run up to the saturation is related to one point of such data, that is, the amount of sorbate retained up to saturation plotted against the inlet concentration (Barros et al., 2004). The dynamic isotherm has the advantage of controlling the fluid-phase concentration. Results of the column dynamic simulations depend on the selection of an appropriate mathematical relation used to represent the equilibrium. Therefore, in order to model the dynamic

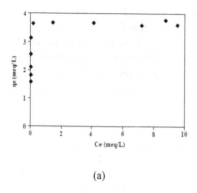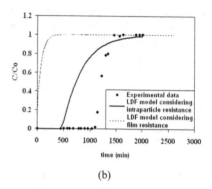

(a)   (b)

Figure 7. a) Batch isotherm of Cr (III) in zeolite NaX at 30 °C (Barros et al., 2004); b) Use of batch isotherm to model Cr (III) breakthrough data in NaX columns

sorption, dynamic isotherms should be considered instead of the frequently used batch iso-therms as they better represent equilibrium in fixed bed (Barros et al., 2009).

The dynamic isotherm of Cr (III) in NaX is shown in Figure 8 at the optimal particle size and flow rate previously determined (Barros et al., 2004). Its shape is totally different, as one can see comparing Figures 8 and 7a.

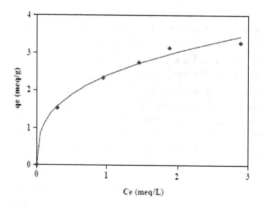

Figure 8. Dynamic isotherm of Cr (III) in zeolite NaX at 30 °C (Barros et al., 2004)

2.6. Modeling of the breakthrough curve

The phenomenological mathematical models are important tools in the design of sorption in fixed bed columns. The validation is done by experimental data obtained in laboratory scale. Mathematical models are useful for designing and optimizing purposes in industrial scale.

The effects of mass transfer in breakthrough curves are complex, mainly if the isotherms are expressed by non-linear mathematical equations. The concentration of the sorbate varies with the position and time. Therefore, the phenomenological model is represented by partial differential equations, which are difficult to be solved analytically.

Due to such complexity, the phenomenological mathematical models have many simplifications to make them feasible to obtain an analytical solution. Bohart and Adams (1920) developed one of the first mathematical models. The respective equation is given by Equation 12. This model was firstly used to describe the dynamic sorption of chlorine in columns packed with activated carbon and are still widely used by several researchers (Singh et al., 2012; Kumar et al., 2011; Rao et al., 2011; Trgo et al., 2011, Chu et al., 2011; Martin-Lara et al., 2010, Muhammad et al., 2010, Borba et al., 2008.). The model Bohart Adams considers that the limiting step of the mass transfer is controlled by the kinetics of sorption which is represented by a second-order reaction.

$$\frac{C_{out}}{C^F} = \begin{cases} 0 & t < t_F \\ \dfrac{1}{\left(e^A + e^{-B} - 1\right)e^B} & t > t_F \end{cases} \tag{12}$$

where: $A = \dfrac{L\ \beta}{u_0}$, $B = \dfrac{(-tu_0 + L)\beta}{\alpha u_0}$, $\alpha = \dfrac{\rho_{bed}q^*}{C^F \varepsilon}$, $\beta = k_a C^F \alpha$ and $t_F = \dfrac{L}{u_0}$

Bohart and Adams model presented only one adjustable parameter that is the rate constant k_a. The parameter q^* has been also reported as an adjustable parameter. However, its value can be determined is the saturation point of the breakthrough curves.

The model developed by Thomas (1944) has also analytical solution. The sorption rate is described by Langmuir adsorption kinetics. Both models, Thomas and Bohart and Adams consider negligible film and particle resistances as well as the dispersion within the column.

The Thomas mathematical model is obtained by mass balances of the sorbate in the fluid phase and in the solid phase. Such mass balances originate an equation that describe the equilibrium in the system. The mass balance in the fluid phase results in the following equation:

$$\frac{\partial C}{\partial t} + \rho_{bed}\frac{(1-\varepsilon)}{\varepsilon}\frac{\partial q}{\partial t} = -u_0\frac{\partial C_j}{\partial z} + D_L\frac{\partial^2 C_j}{\partial z^2} \tag{13}$$

with the following initial and boundary conditions:

$$C(0,z) = 0 \tag{14}$$

In the inlet sample in the column ($z = 0$):

$$D_L \frac{\partial C}{\partial z} = u_0 \left(C(t,0) - C^F \right) \tag{15}$$

In the outlet sample in the column($z = L$):

$$\frac{\partial C}{\partial z} = 0 \tag{16}$$

A rigorous mathematical model that takes into account the variation of the concentration of the adsorbate within the particle is described by the Fick's second law. In these cases, solution of equations of the bed and the particle should be solved simultaneously, which increases the complexity and computational effort. One alternative for reducing the computational effort is to replace Fick's Law by the simplified kinetics equation. (Hsuen, 2000). The approach mostly used to replace the Fick's Law is the LDF model (Linear Driving Force). It applies the expression of the first order kinetics represented by Equation (17). Several authors have used such models with successful results for sorbents and biosorbents.

$$\frac{\partial q}{\partial t} = -K_S \left(q - q_{eq} \right) \tag{17}$$

In the adsorption models represented by Equation (17), it is assumed that the driving force for mass transfer is linear with the concentration of the sorbate in the solid. Moreover, it means the equilibrium condition at the interface between the local phase fluid, as illustrated in Figure 9. The equilibrium in the adsorption process is usually represented by adsorption isotherms, such as: Langmuir, Freundlich, Tóth, Sips. Figure 9 also illustrates the mechanism of the external mass transfer occurs around the surface of the particle. From the interface solid-fluid, the mass transfer occurs, which is represented by following equation:

$$-K_S \left(q - q^* \right) = \frac{K_F a \varepsilon}{\rho_{bed}} \left(C - C_{eq} \right) \tag{18}$$

The mathematical model that considers both mass transfer resistances (external and intraparticle) is called double resistance model. This model is composed by the set of Equations (13)-(18). When the mass transfer in predominantly intraparticle, that is($C^{eq} \approx C$), the equilibrium concentration of the sorbate in the solid is directly related to the concentration in bulk phase. In cases where the mass transfer resistance is in the film, that is $\left(C^{eq} \approx C \right)$, the rate of sorption can be expressed by the following below:

$$\frac{\partial q}{\partial t} = \frac{K_F a \varepsilon}{\rho_{bed}} \left(C - C_{eq} \right) \tag{19}$$

When the operational condition is experimentally optimized, the film and intraparticle resistances of the sorption processes (adsorption, ion exchange or adsorption + ion exchange) are minimized in the experimental range investigated. It means that the film thickness is the thinnest one and there is no significant steric resistance of the sorbate in the particle pores.

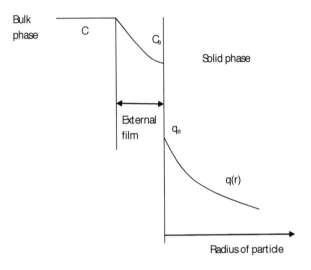

Figure 9. Mass transfer in adsorption process

Some parameters of the model can be calculated from the correlation, namely: axial dispersion coefficient (D_L) and external mass transfer coefficient. The relationship for calculating the axial dispersion is usually expressed from the dimensionless groups: Reynolds (Re), Peclet (Pe), Schmidt (Sc). Delgado (2006) presents several correlations for calculating of axial dispersion of gases and liquids in packed beds. There are many correlations for calculating external mass transfer coefficients in porous media, usually expressed in terms of Reynolds (Re_p), Schmidt and Sherwood (Sh_p). Wakao and Funazkri (1978) developed the following expression for $3 < Re_p < 2000$:

$$Sh_p = 1 + 1.1 Re_p^{0.6} Sc^{1/3} \tag{20}$$

Where: $Sh_p = \dfrac{K_F d_p}{D}$, $Re_p = \dfrac{u_0 d_p}{v}$, $Sc = \dfrac{v}{D}$

The mass transfer coefficient in the solid (K_S) is an adjustable parameter of the model obtained from the experimental breakthrough curves.

2.6.1. Examples of breakthrough modeling

Example 1 – Modeling of Breakthrough curves

Table 2 shows the experimental data of breakthrough curve of zinc in fixed bed columns using Na-Y zeolite as adsorbent obtained by Ostroski (2007) and Table 3 shows the operational conditions and bed parameters.

Time (min)	C_{out}/C^F
0.00	0.0000
22.00	0.0000
31.00	0.0000
55.00	0.0050
68.00	0.0156
74.00	0.1980
87.00	0.3564
93.00	0.5916
118.00	0.7721
136.00	0.8500
147.00	0.9248
180.00	0.9490
205.00	0.9655
230.00	0.9788
250.00	0.9874
260.00	0.9913
280.00	0.9963
307.00	1.0000

Table 2. Experimental data of zinc breakthrough curve

C^f - Feed concentration of adsorbate (meq / cm³)	2.447×10^{-3}
L –Height of the bed (cm)	3.0
m_s - Weight of the adsorbent (g)	0.8
u_0 – Interstitial velocity (cm/min)	25.104
ε - Bed porosity	0.5
ρ_{bed} - Bed density (g / cm³)	0.4192

Table 3. Operational conditions and bed parameters.

From these data it will be tested the Bohart and Adams and the LDF models that consider intraparticle resistance only. Both models have only one adjustable parameter: (k_a) for Bohart and Adams and (K_s) for LDF model.

Bohart and Adams model

From the experimental breakthrough curve it is calculated the capacity of the adsorbent (q^*), since the column experiments are carried outup to the complete exhaustion of the column (saturation point), ie, when the concentration of sorbate in the outlet of the column is equal to the feed concentration. From a mass balance in the column, one can obtain the following equation:

$$q^* = \frac{C^F \dot{Q}}{1000\, m_s} \int_0^{t_{END}} \left(1 - \frac{C_{out}}{C^F}\right) dt \tag{21}$$

Where:

q^*-Capacity of the adsorbent(meq/g);

C_{out}-Concentration of zinc in the outlet of the column (meq/L);

C^F-Feed concentration of zinc (meq/L);

$\dot{Q}$-Volumetric flow rate (cm³/min);

t-time (min);

m_s-Weight of adsorbent (g).

The capacity of the adsorbent was calculated from the area determinate by term $\left(1 - \frac{C_{out}}{C^F}\right)$ calculate by experimental breakthrough curve, illustrated in Figure 10. The integral $\int_0^{t_{END}} \left(1 - \frac{C_{out}}{C^F}\right) dt = 102.78 \text{min}$, calculated by trapeze method.

$$q^* = \frac{C^F \dot{Q}}{1000\, m_s} \int_0^{t_{END}} \left(1 - \frac{C_{out}}{C^F}\right) dt = \left(\frac{2.447 \times 8}{1000 \times 0.8}\right) \times 102.78 = 2.515 \text{ meq/g}$$

$$\alpha = \frac{\rho_{bed}\, q^*}{C^F \varepsilon} = \left(\frac{0.4192 \times 2.515}{2.447 \times 10^{-3} \times 0.5}\right) = 861.7 \text{ meq/g}$$

$$\beta = k_a C_0 \alpha = k_a \times 861.7 \times 10^{-3} \times 2.447 = 2.1086\, k_a$$

$$A = \frac{L}{u_0} \beta = \frac{0.8 \times 2.1086 k_a}{25.104} = 67.20\, k_a$$

$$B = \frac{(-t u_0 + L)\beta}{\alpha\, u_0} = \frac{(-t \times 25.104 + 3) x 2.1086\, k_a}{861.7 \times 25.104} = 9.75 \times 10^{-5} \times (3 - t \times 25.104)\, k_a \text{ min}$$

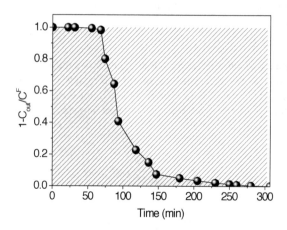

Figure 10. Area of curve $\left(1 - \dfrac{C_{out}}{C^F}\right)$

Applying in the Equation (12), it follows that: $t_F = \dfrac{L}{u_o} = \dfrac{0.8}{25.104} = 0.03187$

The parameter (k_a) estimated from the experimental data of breakthrough curve and applied method for one-dimensional optimization golden search, minimizing the following objective function:

$$F = \sum_{j=1}^{n\_exp} \left(\left(\frac{C_{out}}{C^F}\right)_j^{EXP} - \left(\frac{C_{out}}{C^F}\right)_j^{EXP} \right)^2 \tag{22}$$

Where:

$F = \sum_{j=1}^{n\_exp} \left(\left(\frac{C_{out}}{C^F}\right)_j^{EXP} - \left(\frac{C_{out}}{C^F}\right)_j^{EXP} \right)^2$ -Ratio of the concentrations in the outlet of the column and feed determined experimentally

$\left(\dfrac{C_{out}}{C^F}\right)_j^{EXP}$ -Ratio of the concentrations in the outlet of the column and feed determined by model

$\left(\dfrac{C_{out}}{C^F}\right)_j^{MOD}$ -Number of experimental data

The value of adjustable parameter k_a was 22.35 (meq / cm³ min).

Intraparticle model

In this model considerate that the internal diffusion is controller the mass transfer, the LDF approximation was used to represent the diffusion, described by Equation (17). This model also has only one adjustable parameter (K_S). To solve the equations of the model is also necessary to define an equation for the isotherm and the coefficient of axial dispersion. The Langmuir isotherm, represented by Equation (23), was used in the simulation. The values of parameters used were as follows: D_L = 8.96x10^{-3} cm^2/min, q_{max} = 2.83 meq / g and b = 3.06 L / meq (Ostroski, 2007). The parameter coefficient of mass transfer from adjusted breakhtrough curve was:K_s = 0.0431x10^{-3} min^{-1}.

$$q_{eq} = \frac{q_{max} b C_{eq}}{1 + b C_{eq}} \tag{23}$$

The phenomenological model can provide good results only if the equilibrium model used is appropriate to make this analysis. Its hould compare the values of the equilibrium concentrationsof the adsorbent predicted by the model, in this case calculated by Equation (21) and the experimental value obtained from Equation (23).

Adsorption capacity predicted by the model= $q_{eq} = \frac{q_{max} b\, C_{eq}}{1 + b\, C_{eq}}$ =2.497 meq/g

Experimental adsorption capacity= 2.515 meq/g

In this case, the deviation between predicted adsorption capacity and experimental was 0.7%. It demonstrates that the equilibrium model is appropriated to use in the modeling of the column. When the deviation is high, it is not possible to achieve good adjustments for the breakthrough curves. In these cases, one must first search different isotherm models or theories (RAST –Real Adsorption Solution Theory, VSM - Vacancy Solution Model) to correlate the equilibrium data. In electrolytic systems it can used the Law of Action Mass.

When the parameters of model equilibrium are estimated from batch systems, the model accurately describes the behavior of the equilibrium curve. However, these values are different from the one obtained in the capacity of the column. Such behavior occurs generally for electrolyte systems. Several authors (Bajracharya and Vigneswaran 1990; Silva et al. 2002; Palma et al. 2003; Barros et al 2004; Izquierdo[a,b] et al., 2010) reported the differences between batch and column capacity of adsorption. In these cases for modeling the dynamic adsorption column fixed bed must perform the adjustment of the parameters from the equilibrium data obtained in fixed bed column at different feed concentration. See section 2.5.

To solve the partial differential equations (Eq. 13 and Eq.17) it was used the lines method, which transforms the partial differential equations in a system of ordinary differential equations by approximating the spatial derivatives by finite differences. Further details of the resolution method used can be found in Silva (2001). The results obtained with the LDF model (Numerical solution) and Adams Bohart (analytical solution) is shown in Figure (11).

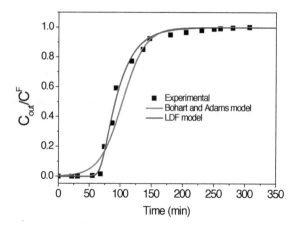

Figure 11. Zinc Breakthroughs curves

In this example, the model LDF described more accurately the behavior of zinc break-through curve due the consideration of mass transfer controlled by the intraparticle diffusion. It is more realistic than consideration that the kinetics is phase controller, which is the hypothesis of the Bohart and Adams model.

3. Final comments

Design of dynamic sorption is considered a simple process. Nevertheless, to reach higher and higher efficiencies is always a challenge.

Differences in packing procedures may provide quite different breakthrough behavior. Therefore, much attention should be paid to avoid channeling and bubbles of different fluid phase. Channeling promotes undesired dispersion in the bed keeping the system away from plug-flow behavior. Bubbles create additional mass transfer resistances and diminish the uptake efficiency. Much attention should be paid in the operational conditions imposed to the column. The ones that minimize the mass transfer resistances are always recommended.

Mathematical modeling of dynamic behavior with analytical solutions, such as: Bohart and Adams and Thomas considers that the limiting step is controlled by adsorption kinetics and can be applied only to monocomponent systems. The advantage of these models is that the estimation is simpler. Models with a numerical solution is more realistic because they take into account various aspects related to mass transfer (axial dispersion, external film diffusion and intraparticle diffusion) are more suitable for use in the design and optimization of sorption processes in industrial scale. The phenomenological models are generated by mass balances of adsorbate in fluid and solid phase and require information of mass transfer mechanism and equilibrium. A prerequisite for phenomenological models is the theory and / or model used to represent the equilibrium of adsorption (isotherms, the law of mass action, RAST, VSM). Particularly in electrolyte systems, it is recommended that the equilibrium data are obtained in columns experiments on the basis of no pH correction. As already seen, the equilibrium conditions are different in batch and in dynamic systems. In the batch system, the counter ions are in solution, whereas in systems in a fixed bed column is loaded by the feed solution. The LDF model is an approximation to represent the intraparticle diffusion and has been shown to be efficient to describe the behavior of breakthrough curves for different systems (monocomponent and mixtures).

Nomenclature

a-Mass transfer area per unit bed volume (m^{-1})

$\dfrac{2.83 \times 3.06 \times 2.447}{1 + 3.06 \times 2.447}$-Concentration of the adsorbate in the fluid phase (mmol / L);

C-Feed concentration of the adsorbate in the fluid phase (mmol / L);

C^F-Equilibrium concentration of the adsorbate fluid phase (mmol / L);

C_{eq}-Particle diameter (m);

d_p-Molecular coefficient diffusion (m^2/min);

D-Axial dispersion coefficient (m^2/min);

D_L -Height of bed (m);

L -Mass transfer coefficient in the solid (min^{-1})

k_a-Kinetics parameters of Bohart and Adams model (L / mol min)

K_S-External film mass transfer coefficient (m/mim^{-1})

u_0-Interstitial velocity (m/min);

K_F-Coordinated in the axial direction (m);

z-Bed porosity;

ε-Bed density (g / L);

ρ_{bed}-Kinematic viscosity (m^2/min);

Author details

M. A. S. D. Barros[1], P. A. Arroyo[1] and E. A. Silva[2]

1 Department of Chemical Engineering, State University of Maringá, Maringá, Brazil

2 Department of Chemical Engineering, West Paraná State University, Jardim La Salle, Toledo, Brazil

References

[1] Bailey, J.E.,Ollis, D.F.:Biochemical engineering fundamentals. McGraw-Hill, New
York, (1986).

[2] Bajracharya, K., Vigneswaran, S., Adsorption of cadmium and zinc in saturated soil
columns: Mathematical models and experiments, Environmental Technology, 11,
9-24. 1990.

[3] Barros, M,A,S,D, Zola, A,S., Arroyo, P.A., Sousa-Aguiar, E.F., Tavares, C.R.G. Binary
Ion Exchange of Metal Ions in Y and X Zeolites, Brazilian Journal of Chemical Engineering, 20, 4, 413-421, 2003.

[4] Barros, M.A.S.D., Silva, E.A., Arroyo, P.A., Tavares, C.R.G., Schneider, R.M., Suszek,
M., Sousa-Aguiar, E.F., Removal of Cr(III) in the fixed bed column and batch reactors
using as adsorbent zeolite NaX, Chemical Engineering Science, 59, 5959 – 5966, 2004.

[5] Barros, M.A.S.D., Zola, A.S, Tavares, C.R.G., Sousa-Aguiar, E.F. Chromium uptake
from tricomponent solution in zeolite fixed bed, Adsorption, 12, 229-248, 2006.

[6] Barros, M.A.S.D., Zola, A.S., Arroyo, P.A., Sousa-Aguiar, E.F., Tavares, C.R.G.Equili-
brium and dynamic ion exchange studies of Cr3+ on zeolites NaA and NaX, Acta Sci-
entiarum, 24, 6, 1619-1625, 2002.

[7] Bohart, G. S., Adams, E. Q., Some aspects of the behavior of charcoal with respect to
chlorine,Journal of the American Chemical Society, 42, 523-544, 1920.

[8] Borba, C. E., Silva, E.A., Fagundes-Klen, M. R., Kroumov, A. D., Guirardello, R. , Prediction of the copper (II) ions dynamic removal from a medium by using mathematical models with analytical solution,Journal of Hazardous Materials, 152 (1), 366-372,
2008.

[9] Chu, K. H., Kim, E. Y., Feng, X., Batch Kinetics of Metal Biosorption: Application of
the Bohart-Adams Rate Law, Separation Science and Technology, 46 (10), 1591-1601,
2011.

[10] Comiti, J., Mauret, E., Renaud, M., Mass transfer in fixed beds: proposition of a generalizedcorrelation based on an energetic criterion, Chemical Engineering Science, 55, 5545-5554, 2000.

[11] Delgado, J.M.P.Q., A critical review of dispersion in packed bed, Heat Mass Transfer, 42, 279-310, 2006.

[12] Fogler, H.S., Elements of Chemical Reaction Engineering, 3rd ed., Prentice Hall, New Delhi- India, 2004.

[13] Gazola, F. C., Pereira, M.R., Barros, M.A.S.D, Silva, E.A, Arroyo, P.A. Removal of Cr3+ in fixed bed using zeolite NaY, Chemical Engineering Journal, 117, 253–261, 2006.

[14] Geankoplis, C.J., Transport Processes andUnit Operations, 3rd ed., PTR Prentice Hall, USA, 1993.

[15] Helferich, F.,Ion Exchange, Dover Publications Inc., New York, 1995.

[16] Hill, C.G.,An Introduction to Chemical Engineering Kinetics and Reactor Design, John Wiley & Sons, USA, 1977.

[17] Hsuen, H. K., An improved linear driving force approximation for intraparticle adsorption, Chemical Engineering Science, 55,3475-3480, 2000.

[18] Izquierdoa, M., Gabaldón, C., Marza, P., Álvarez-Hornos, F. J., Modeling of copper fixed-bed biosorption from wastewater by Posidonia oceanica, Bioresource Technology, 101, 510–517, 2010.

[19] Izquierdob, M., Gabaldón, C., Marzal, P., Sempere, F., Sorption of copper by a highly mineralized peat in batch and packed-bed systems, Journal of Chemical Technology and Biotechnology, 85, 165-172, 2010.

[20] Jain, J.S., Snoeyink, V.L.: Adsorption from biosolute systems on active carbon. J. Water Pollut. Control Fed. 45, 2463-2479 (1973).

[21] Kang, S. Y., Lee, J. U., Moon, S. H., Kim, K. W., 2004, Competitive adsorption characteristics of Co2+ Ni2+ and Cr3+ by IRN 77 cation exchange resin in synthesized wastewater, Chemosphere, v. 56, pp.141–147.

[22] KLamer, K, Van Krevelen, D. W., Studies on ion Exchange-I, Chem. Eng. Sci, 7, 4, 197-203, 1958.

[23] Klein, G., Tondeur, D., Vermeulen, T. Multicomponent ion exchange in fixed beds, Ind. Eng. Chem. Fundam., 6, 3, 339-351, 1967.

[24] Kratochvil, D., Volesky, B., Demopoulos, G. Optimizing Cu removal/recovery in a biosorption column, Wat. Res., .31, 9, 2327-2339, 1997.

[25] Kumar, J, Chatterjee, A., Schiewer, S., Biosorption of Cadmium(II) Ions by Citrus Peels in a Packed Bed Column: Effect of Process Parameters and Comparison of Different Breakthrough Curve Models,Clean-Soil Air Water, 39 (9),874-881, 2011.

[26] Labanda, J., Sabaté, J., Llorens, J., Experimental and modeling study of the adsorption of single and binary dye solutions with an ion-exchange membrane adsorber, Chemical Engineering Journal, 166, 536–543, 2011.

[27] Martin-Lara, M. A., Hernainz, F., Blazquez, G., Tenorio, G., Calero, M., Sorption of Cr (VI) onto Olive Stone in a Packed Bed Column: Prediction of Kinetic Parameters and Breakthrough Curves, Journal of Environmental Engineering-ASCE, 136 (12), 1389-1397, 2010.

[28] McCabe, W. L., Smith, J.C., and Harriot, P.Unit Operations of Chemical Engineering, McGraw-Hill International Ed., 6th ed., New York, USA, 2001.

[29] Misak, N.Z., Some aspects of the application of adsorption isotherms to ion exchange reactions. React. Funct. Polym., 43, 153-164 (2000).

[30] Mohan, D. and Chander, S., Single Component and Multi-component Adsorption of Metal Ions by Activated Carbons, Colloids and Surfaces A: Physicochemical and Engineering Aspects, 177, 2-3, 183-196, 2001.

[31] Muhamad, H., Doan, H., LOHI, A., Batch and continuous fixed-bed column biosorption of Cd2+ and Cu2+, Chemical Engineering Journal, 158 (3), 369-377, 2010.

[32] Myers, A. L.; Prausnitz, J. M.; Thermodynamics of mixed-gas adsorption; AIche Journal, 11, 1, 121-127, 1965.

[33] Ostroski, I. C., Borba, C. E., Silva, E. A., Arroyo, P. A., Guirardello, R., Barros, M. A.S.D. Mass Transfer Mechanism of Ion Exchange in Fixed Bed Columns, J. Chem. Eng. Data, 56, 375-382, 2011.

[34] Ostroski, I.C., Mechanism of removal of Iron(III) and Zinc(II) ontoZeolite Na-Y, In Portuguese, Master Thesis, UEM - State University of Maringá, Brazil, Chemical Engineering Department , 2007.

[35] Ostroski, I.C., Barros, M.A.S.D. , Silva, E. A., Dantas, J.H., Arroyo, P.A., Lima, O.C.M., A comparative study for the ion exchange of Fe(III) and Zn(II) on zeolite NaY, Journal of Hazardous Materials, 161 , 1404–1412, 2009.

[36] Palma, G; Freer, J; Baeza, J.; Removal of metal ions by modified Pinus radiata bark and tannins from water solutions, Water Research, 37, 4974-80, 2003.

[37] Pereira, M. R., Arroyo, P.A., Barros, M.A.S.D., Sanches, V.M., Silva, E.A., Fonseca, I. M., Lovera, R.G. Chromium adsorption in olive stone activated carbon, Adsorption, 12, 155–162, 2006.

[38] Perry, R. H., Perry's Chemical Engineers' Handbook, 7th. Ed.,Mc-Graw Hill, 1999.

[39] Rao, K. S., Anand, S., Venkateswarlu, P., Modeling the kinetics of Cd(II) adsorption on Syzygium cumini L leaf powder in a fixed bed mini column,Journal of Industrial and Engineering Chemistry, 17 (2), 174-181, 2011.

[40] Sağ, Y., I. Ataçoğlu, and T. Kutsal, Equilibrium Parameters for the Single and Multi-component Biosorption of Cr(VI) and Fe(III) Ions on R. Arrhizus in a Packed Column, Hidrometallurgy, 55, 165–179, 2000.

[41] Sánchez, A., Ballester, A., Blásquez, M.L., González, F., Muñoz, J., Hammaini, A.: Biosorption of copper and zinc by cymodocea nodosa. FEMS Microbiol. Rev. 23 527-536 (1999).

[42] Silva, E.A., Biosorption of chromium(III) and copper (II) ions on seaweed Sargassum sp. in fixed bed columns. In Portuguese, Doctorate thesis, UNICAMP, University State of Campinas, Brazil,Faculty ofChemical Engineering,2001.

[43] Silva, E.A., Cossich, E.S., Tavares, C.R.G., Filho, L.C., Guirardello, R. Modeling of copper(II) biosorption by marine alga Sargassum sp. In fixed-bed column, Process Biochemistry, 38, 791-799, 2002.

[44] Singh, A., Kumar, D., Gaur, J. P., Continuous metal removal from solution and industrial effluents using Spirogyra biomass-packed column reactor,Water Research, 46 (3), 779-788, 2012.

[45] Sprynskyy, M., Buszewski, B., Terzyk, A.P., Namiesnik, K.: Study of the selection mechanism of heavy metal (Pb2+, Cu2+, Ni2+ and Cd2+) adsorption on clinoptolilite. J. Colloid Interface Sci. 304, 21-28 (2006).

[46] Suzuki, M. Adsorption Engineering, Elsevier Science Publishers B.V., 1990.

[47] Thomas, H.C., Heterogeneous ion exchange in a flowing system, Journal of the American Chemical Society, 66, 1664–1666, 1944.

[48] Trgo, M., Medvidovic, N. V., Peric, J., Application of mathematical empirical models to dynamic removal of lead on natural zeolite clinoptilolite in a fixed bed column, Indian Journal of Chemical Technology, 18 (2), 123-131, 2011.

[49] Valdman, E., Erijman, L., Pessoa, F.L.P. and Leite, S.G.F. Continuous Biosorption of Cu and Zn by Immobilized Waste Biomass Sargassum sp., Process Biochemistry, 36, 869-873, 2001.

[50] Walker, G.M., Weatherly, L.R. Adsorption of acid dyes on to granular activated carbon in fixed beds, Water Research, 31, 8, 2093–2101, 1997.

Chapter 8

Subcritical Water Extraction

A. Haghighi Asl and M. Khajenoori

Additional information is available at the end of the chapter

1. Introduction

Extraction always involves a chemical mass transfer from one phase to another. The principles of extraction are used to advantage in everyday life, for example in making juices, coffee and others. To reduce the use of organic solvent and improve the extraction methods of constituents of plant materials, new methods such as microwave assisted extraction (MAE), supercritical fluid extraction (SFE), accelerated solvent extraction (ASE) or pressurized liquid extraction (PLE) and subcritical water extraction (SWE), also called superheated water extraction or pressurized hot water extraction (PHWE), have been introduced [1-3].

SWE is a new and powerful technique at temperatures between 100 and 374°C and pressure high enough to maintain the liquid state (Fig.1) [4]. Unique properties of water are namely its disproportionately high boiling point for its mass, a high dielectric constant and high polarity [4]. As the temperature rises, there is a marked and systematic decrease in permittivity, an increase in the diffusion rate and a decrease in the viscosity and surface tension. In consequence, more polar target materials with high solubility in water at ambient conditions are extracted most efficiently at lower temperatures, whereas moderately polar and non-polar targets require a less polar medium induced by elevated temperature [5].

Based on the research works published in the recent years, it has been shown that the SWE is cleaner, faster and cheaper than the conventional extraction methods. The essential oil of Z. *multiflora* was extracted by SWE and compared with two conventional methods, including hydrodistillation and Soxhlet extraction [7]. The total extraction yields found for the total essential oil of Z. *multiflora* were 2.58, 1.51 and 2.21% (w/w) based on the dry weight for SWE, hydrodistillation and Soxhlet extraction, respectively.

The comparison among the amount of thymol and carvacrol (milligram per gram dried sample) by SWE, hydrodistillation and Soxhlet extraction is shown in Table 1 [7]. The amount of valuable oxygenated components in the SWE method is significantly higher than hydro-

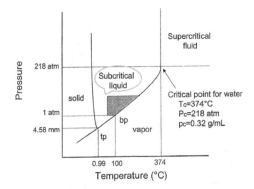

Figure 1. Phase diagram of water as a function of temperature and pressure (Cross-hatched area indicates the preferred region (SWE)) [4].

Components	SWE*	Hydrodistillation†	Soxhlet extraction‡	RI§
Thymol	9.25 (4.77%)**	4.38 (2.97%)	0.94 (2.78%)	1,232
Carvacrol	11.51 (4.33%)	4.06 (3.31%)	1.39 (2.83%)	1,242

Sample weight = 4 g; particle size = 0.5 mm; flow rate = 2 mL/min; temperature = 150C; pressure = 20 bar; and extraction time = 150 min.
* Extraction time = 150 min.
** Relative SD percent.
† Extraction time = 180 min.
‡ Extraction time = 210 min.
§ Retention indices (RI) on the DB-5 column.

* Extraction time = 150 min.

** Relative SD percent.

1' Extraction time = 180 min.

i Extraction time = 210 min.

§ Retention indices (RI) on the DB-5 column.

Table 1. The amount of Thymol and carvacrol (mg/g dried sample) of the essential oil of *Z.multiflora*, extracted by SWE, hydrodistillation and Soxhlet extraction [7].

distillation and Soxhlet extraction. As hexane is a nonpolar solvent, non-oxygenated components are enhanced compared to subcritical water. On the other hand, in general, non-oxygenated components present lower vapor pressures compared to oxygenated components, and in this sense, its content in hydrodistilled extracts are increased. Because of the significant presence of the oxygenated components, the final extract using the SWE method was relatively better and more valuable.

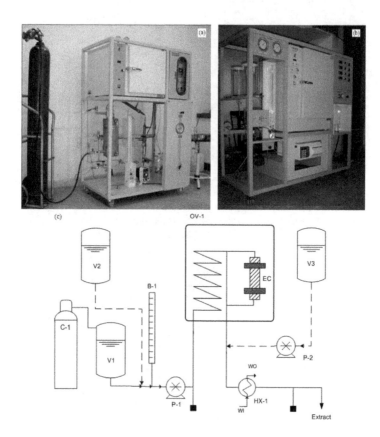

Figure 2. (a) and (b) pictures of first and second generated systems respectively and (c) Schematic diagram of SWE system, B-1: Burette, C-1: Nitrogen cylinder, EC: Extraction cell, HX-1: Heat Exchanger, OV-1: Oven, P-1, 2: Pumps, V-1: Water tank, V-2: Solvent tank, V-3: Rinsing solvent tank, WI: Water inlet, WO: Water outlet.

2. Equipment of subcritical water extraction

No commercial SWE equipment is available, but the apparatus is easy to construct in the laboratory. SWE is performed in batch or continue systems but continue system is current. In this system, extraction bed is fixed and flow direction is usually up to down for easily cleaning of analytes. Already, we worked SWE of Z. *multiflora* with first generated system (Iranian Research Organization for Science and Technology [IROST], Tehran, Iran) [7] (Fig. 2(a)). Afterit, laboratory-built apparatus of SWE equipment was designed by studying of different systems and have advantages versus first generation (Semnan University, Semnan, Iran) (Fig. 2(b)).

The second generated system is presented in Fig. 2(c). The main parts of a dynamic SWE unit are the following: three tanks, two pumps, extraction vessel, oven for the heating of the

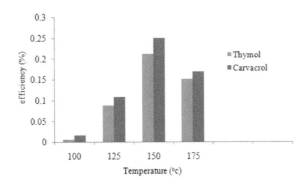

Figure 3. Effect of temperature on the main essential oil components SWE of *Z. multiflora*. Operating conditions: sample weight=4.0 g; flow rate=2 mL/min; particle size=0.50 mm; pressure=20 bar; and extraction time=60 min.

extraction vessel, heat Exchanger for cooling of exteract, pressure restrictor and sample collection system. One of the pumps is employed for pumping the water (and extract) and another pump is employed for flushing the tubings. Also, one of tank is employed for organic solvent that is as a main solvent or co-solvent. A pressure restrictor is needed to maintain the appropriate pressure in the equipment. It was constructed of stainless steel.

3. Effective parameters in subcritical water extraction

3.1. Effect of temperature

One of the most important parameters affecting SWE efficiencies is the extraction temperature. As the temperature rises, there is a marked and systematic decrease in permittivity, an increase in the diffusion rate and a decrease in the viscosity and surface tension. SWE must be carried out at the highest permitted temperature. It should be mentioned that increasing the extraction temperature above a certain value gives rise to the degradation of the essential oil components. The maximum permitted extraction temperature must be obtained experimentally for different plant materials. Regarding the extraction of essential oils, it has been shown that temperatures between 125 and 175°C will be the best condition. The extraction temperature for *Z. multiflora* was optimized in order to maximize the efficiency of thymol and carvacrol (structural isomers) as key components (more than 72%) [7]. Its influence was studied between 100 and 175°C, and the mean particle size, flow rate, extraction time and pressure were selected to be 0.5 mm, 2 ml/min, 60 min and 20 bar pressure, respectively (Fig. 3).

It was seen that the efficiency of thymol and carvacrol increased generally with increase in temperature up to 150°C. At 175°C, it decreased, and an extract with a burning smell was produced. It may be the result of degradation of some of the constituents at higher temperatures. Because of the highest efficiency of thymol and carvacrol essential oil at 150°C and the

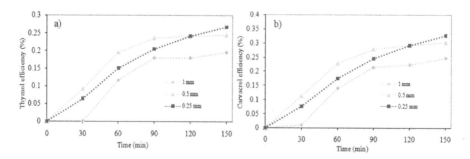

Figure 4. Effect of particle size on the efficiency of a) thymol and b) carvacrol SWE of *Z.multiflora*. Operating conditions: sample weight = 4.0g; flow rate = 2 ml/min; temperature = 150°C; and pressure = 20 bar [7].

disagreeable odor of the extract at higher temperatures, further experiments were carried out at this temperature.

3.2. Effect of particle size

The effect of the mean particle size on the efficiency of thymol and carvacrol at 150°C temperature, 2ml/min flow rate, 20 bar pressure and 150 min extraction time is shown in Fig. 4 [7]. The mean ground leaf particles were selected to be 0.25, 0.5 and 1.0 mm. The final amount of thymol and carvacrol extracted from 0.5-mm-size particles was near to 0.25-mm particles. It shows that, at least in the selected range of mean particle sizes (0.25-0.5 mm), the extraction process may not be controlled by the mass transfer of thymol and carvacrol. It was expected that the rate of the 0.25-mm-size particles was more than the 0.5-mm-size particles, but it did not happen.

A possible explanation for this observation could be that the particles were close fitting at initial times, and the extraction was done slowly. After the expired time, the close fitting particles opened from each other, and at final extraction value of the 0.25-mm-size particles was more than the 0.5-mm-size particles. Regarding the larger 1.0-mm-size particles, the efficiency is substantially lower. It shows that the process may be controlled by the mass transfer of thymol and carvacrol for larger particle sizes.

To prevent the probable vaporization of the essential oils during the grinding of the leaves and also to make the work of the filters easier, for further experiments, the best value of mean particle size was selected as 0.50 mm.

3.3. Effect of flow rate

The effect of water flow rate on the efficiency of thymol and carvacrol at 150°C temperature, 0.5-mm-particle size, 20 bar pressure and 150 min extraction time is shown in Fig. 5 [7]. The water flow rate was studied at 1, 2 and 4 ml/min. As can be seen, the rate of the essential oil extraction was faster at the higher flow rate. The rate is slower at 2 ml/min and even slower at 1 ml/min. It is in accordance with previous works [8, 9]. It means that the mass transfer of the

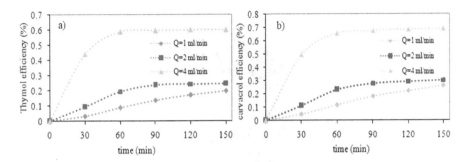

Figure 5. Effect of flow rate on the efficiency of a) thymol and b) carvacrol SWE of *Z. multiflora*. Operating conditions: sample weight = 4.0g; particle size= 0.5 mm; temperature = 150°C; and pressure = 20 bar [7].

thymol and carvacrol components from the surface of the solid phase into the water phase regulated most of the extraction process. Increase in flow rate resulted in increase in superficial velocity, and thus, quicker mass transfer [10]. The main disadvantage of applying higher water flow rates is increasing the extract volume and consequently, lower concentration of the final extracts. In practice, the best flow rate must be selected considering two important factors, including the extraction time and the final extract concentration. It is clear that a shorter extraction time and more concentrated extracts are desirable. To prevent a slower extraction rate and longer extraction times, despite the larger amount of the final extracts, a flow rate of 2 ml/min was selected as the optimum value.

While temperature, particle size and flow rate extraction are the main parameters affecting SWE, type of analyte, extraction vessel characteristics and use of modifiers and additives are also important. Although matrix and other effects play a role, many of these are less critical in SWE than in SFE because of the harsh extraction conditions (high temperature) typical in SWE, particularly for non-polar analytes.

4. Extraction mechanism

The SWE process can be proposed to have six sequential steps: (1) rapid fluid entry; (2) desorption of solutes from matrix active sites; (3) diffusion of solutes through organic materials; (4) diffusion of solutes through static fluid in porous materials; (5) diffusion of solutes through layer of stagnant fluid outside particles; and (6) elution of solutes by the flowing bulk of fluid (Fig. 6).

As we know, the extraction rate is limited by the slowest of these three steps. The effect of step (1) is typically small and often neglected. Although the diffusion of the dissolved solute within the solid is usually the rate limiting step for most botanicals, partitioning of solute between the solid matrix and solvent have been reported as the rate-limiting mechanism for SWE of essential oil from savory [10].

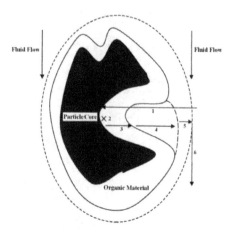

Figure 6. Proposed schematic presentation of the extraction steps in SWE.

The plots the amount of compound extracted versus solvent flow rates and versus solvent volume can determine the relative importance of these steps. For example, if the rate of extraction is controlled by intra-particle diffusion or kinetic desorption, the increase in bulk fluid flow rate would have little effect on extraction rate. On the other hand, if the extraction is controlled by external film transfer diffusion, extraction rates increase with solvent flow rate. In the case where the extraction rate is controlled by thermodynamic partitioning, doubling the bulk fluid flow rate would double the extraction rate, while the curves of extraction efficiency versus the volume of water passed for all flow rates would overlap. In one of our previous work, four proposed models have been applied to describe the extraction mechanisms obtained with SWE of *Z. multiflora* essential oil. These were included (1) partitioning coefficient model, (2) one-site (3) two-site desorption models and (4) thermodynamic partition with external mass transfer model [11]. In other studying unsteady state mass balance of the solute in solid and subcritical water phases (two-phase model) was investigated [12]. Also Computational Fluid Dynamics (CFD) modeling of extraction was considered [13].

5. Modeling of SWE

5.1. Thermodynamic model (Partitioning coefficient (K_D) model)

Partitioning coefficient model, adopted from Kubatova et al. [10], describes the extraction process that is controlled by partitioning of solute between matrix and solvent similar to elution of solute from a partition chromatography column. For extraction, this type of behavior occurs when the initial solute concentration in the plant matrix is small. This model assumes that the initial desorption step and the subsequent fluid-matrix partitioning is rapid. Here the thermodynamics partitioning coefficient, K_D, is defined as:

$$K_D = \frac{Concentration\ of\ solute\ in\ the\ matrix}{Concentration\ of\ solute\ in\ the\ extraction\ fluid}\ ; at\ equilibrium \tag{1}$$

Hence the Extraction with subcritical water can be fitted using this simple thermodynamic model. The mass of analyte in each unit mass of extraction fluid and the mass of analyte remaining in the matrix at that period in the entire extraction time is based on the K_D value determined for each compound. The thermodynamic elution of analytes from matrix was the prevailing mechanism in SWE as evidenced by the fact that extraction rate increased proportionally with the subcritical water flow rate. Therefore, if the K_D model applies to a certain extraction, the shape of an extraction curve would be defined by:

$$\frac{M_b}{M_i} = \frac{\left(1 - \frac{M_a}{M_i}\right)}{\left(\frac{K_D m}{(V_b - V_a)\rho} + 1\right)} + \frac{M_a}{M_i} \tag{2}$$

M_a: cumulative mass of the analyte extracted after certain amount of volume V_a (mg/g dry sample)

M_b: cumulative mass of the analyte extracted after certain amount of volume V_b (mg/g dry sample)

M_i: total initial mass of analyte in the matrix (mg/g dry sample)

M_b/M_i and M_a/M_i: cumulative fraction of the analyte extracted by the fluid of the volume V_b and V_a (ml)

K_D: distribution coefficient; concentration in matrix/concentration in fluid

ρ: density of extraction fluid at given condition (mg/ml)

e: exponential function

m: mass of the extracted sample (mg dry sample).

The model eq. (1) and the experimental data for Z. *multiflora* from all volumetric flow rate, were used to determine the K_D value by minimizing the errors between the measured data and the K_D model using Matlab curve fitting solver. The values of K_D are shown in Table 2 for different flow rates [11]. It was demonstrated that individual essential oil compounds have a range of K_D values from ~4 to ~250 [10].

Flow rate/	K_D	
ml·min^{-1}	Thymol	Carvacrol
1	80	70
2	80	70
4	2	2

Table 2. K_D values of partitioning coefficient model for different volumetric flow rates [11].

5.2. Mass transfer models

5.2.1. Diffusion model

Mass transfer can be defined as the migration of a substance through a mixture under the influence of a concentration gradient in order to reach chemical equilibrium. The diffusion coefficient (D_e) is the main parameter in Fick's law, and application of this mathematical model to solid foods during solid-liquid extraction is a common way to calculate the effective diffusion coefficient (Crank, 1975 [14]). However, Gekas (1992) noted, values of D_e can vary by several orders of magnitude for the same material which may be due to structural changes in the food material during different stages of the process [15]. Therefore, it is important to keep a constant particle size as breakage of cell wall or grinding can reduce the particle size and hence decrease the distance for solute to travel from inside to surface of particle.

Fick derived a general conservation equation for one-dimensional non-steady state diffusion when the concentration within the diffusion volume changes with respect to time, known as Fick's second law (Cussler, 1984; Mantell et al., 2002) [16, 17]:

$$\frac{\partial C}{\partial t} = D_e \frac{\partial^2 C}{\partial r^2}$$

With the initial condition:

$$C_{(t=0)} = C_o \qquad\qquad\qquad \text{a}$$

And boundary conditions: $\qquad\qquad\qquad\qquad\qquad\qquad\qquad\qquad$ (3)

$$\frac{\partial C}{\partial r}_{(r=0)} = 0 \qquad\qquad\qquad \text{b}$$

$$C_{(r=R)} = 0 \qquad\qquad\qquad \text{c}$$

Where C is the solute concentration (mg/ml) at any location in the particle at time t (s); C_o is the initial solute concentration (mg/ml); D_e is the effective diffusion coefficient (m²/s) assuming that D_e is constant with the concentration; t is extraction time (s); r is the radial distance from the centre of a spherical particle (m); R is radius of spherical particle (m).

Various solutions of Fick's second law have been presented for the diffusion of a compound during solid-liquid extraction depending on the shape of the particle. An approximate numerical solution to Fick's second law (eq. 2) for a spherical particle was given by Crank (1975) and Cussler (1984):

$$\frac{M_t}{M_\infty} = 1 - \frac{6}{\pi^2} \sum_{n=1}^{\infty} \frac{1}{n^2} \exp\left[\frac{-D_e n^2 \pi^2 t}{R^2} \right] \qquad\qquad (4)$$

Where M_t: total amount of solute (mg/g) removed from particle after time t, M_∞: maximum amount (mg/g) of solute extracted after infinite time. M_t/M_∞: ratio of total migration to the maximum migration concentration, R: average radius of an extractable particle. When time becomes large, the limiting form of Eq. 3 becomes:

$$1 - \frac{M_t}{M_\infty} = \frac{6}{\pi^2} \exp\left[\frac{-D_e \, \pi^2 t}{R^2}\right] \tag{5}$$

To determine the effective diffusion coefficient values two methods were used. The first method was a linear (graphical) solution in which D_e was determined from the slope of the ln $(1-M_t/M_\infty)$ vs. time plot (Dibert et al., 1989) [18]. Thus, eq. 4 can be solved by taking the natural logarithm of both sides. It shows that the time to reach a given solute content will be directly proportional to the square of the particle radius and inversely proportional to D_e:

$$ln\left[1 - \frac{M_t}{M_\infty}\right] = \ln \frac{6}{\pi^2} - \frac{-D_e \, \pi^2 t}{R^2} \tag{6}$$

Where slope $= \frac{\pi^2 D_e}{R^2}$.

The second method of solution used involved nonlinear regression with effective diffusivity (D_e) as a fitting parameter. In this method, the effective diffusivity D_e was estimated from eq. 4 using a Microsoft Excel Solver program. The program minimizes the mean square of deviations between the experimental and predicted $\ln(1-M_t/M_\infty)$ values (Tutuncu and Labuza, 1996) [19]. The first 10 terms of the series solution are taken into consideration by the program as the solution to the series becomes stable after 10 terms (n=10).

Most researchers in this area have adopted diffusion models based on solutions to Fick's second law for various defined geometrical shapes of the solid given in Crank method, usually based on infinite or semi-infinite geometries of a slab (plane sheet), a cylinder or a sphere. For example, a slab can be used to describe an apple slice or a sheet of herring muscle; a sphere for the description of coffee beans or particles of cheese curd, and a cylinder for the description of cucumber pickles. There is considerable variation in the solutions adopted by various researchers. The starting point for modeling a particular diffusion process is to consider the shape of the solid and the nature of the process itself: uptake of solute into the food, leaching of solute from the food or diffusion of solute through the food, and the experimental conditions in terms of initial and equilibrium solute concentrations.

The solution models usually consider a uniform initial solute concentration throughout the food, no resistance to mass transfer in the diffusion medium and no chemical reaction; but vary for a particular geometry depending on the solute concentrations at the surface of the solid, the volume of the solution (and therefore the relative change in solute concentration in the external solvent), and the time period of the experiment. A representative selection of solutions is given in Table 3. For example, Bressan et al. investigated solute diffusional loss from coffee beans and cottage cheese curd, respectively [20]; they chose solution models to Fick's second law from Crank which assumed the geometry of the solids approximate to that of infinite spheres. However, the former researchers considered a system in which the solute concentration in the external solvent remains effectively constant and zero throughout the extraction. This can be the case for large infinite solvent volumes and small solute solid concentrations. The latter group of researchers considered a process where the concentration of the solute in the surrounding medium changes significantly. This can be the case for small solvent volumes and high solid phase solute concentrations in the case of leaching, and high external solvent concentrations in the case of solute uptake processes; and consequently adopted a different model.

5.2.2. One-site kinetic desorption model

One-site kinetic desorption model describes the extractions that are controlled by intra-particle diffusion. This occurs when the flow of fluid is fast enough for the concentration of a particular solute to be well below its thermodynamically controlled limit. The one-site kinetic model was derived based on the mass transfer model that is analogous to the hot ball heat transfer model [29, 30]. The assumptions are that the compound is initially uniformly distributed within the matrix and that, as soon as extraction begins, the concentration of compound at the matrix surfaces is zero (corresponding to no solubility limitation). For a spherical matrix of uniform size, the solution for the ratio of the mass, M_r, of the compound that remains in the matrix sphere after extraction time, t, to that of the initial mass of extractable compound, M_i is given as:

$$\frac{M_r}{M_i} = \frac{6}{\pi^2} \sum_{n=1}^{\infty} \frac{1}{n^2} \exp(-D_e n^2 \pi^2 t / r^2) \tag{7}$$

In which n is an integer and D_e is the effective diffusion coefficient of the compound in the material of the sphere (m²/s).

Diffusion process	Diffusion equation	Experimental measurements	Calculation Diffusivity
proteins Diffusion though a potato disk [21]	$M_t = \frac{S D_e C_{L\,1} t}{a} + \frac{2aS}{\pi^2} =$ $1 - \sum_{n=1}^{\infty} \left(C_{L\,1} \cos\left(n\pi\right) \exp\left[\frac{-D_e n^2 t}{a^2}\right] \right)$	Protein concentration on the source side initially; at intervals on receiving side	Fitting the experimental values for M_t to the equation by non-linear regression
Analytes diffusion though an apple disk [22]	$M_t = \frac{X A D_e (C - C_L)}{2a}$ A simple lumped parameter equation model	soluble analytes content in the limited volume of solvent	A single effective diffusivity was calculated for each set of data directly from the formula
Analytes diffusion though cheese curd [19]	$\frac{M_t}{M_\infty} = 1 - \sum_{n=1}^{\infty} \frac{6a(1+a)}{9 + 9a + q_n^2 a^2} \exp\frac{-D_e q_n^2 t}{2a} :$	Samples of solvent surrounding curd	Experimental data fitted to the equation model

Diffusion process	Diffusion equation	Experimental measurements	Calculation Diffusivity
	Solution of Fick's 2nd law for solute loss or uptake for sphere geometry but for finite volume of solvent with attainment of equilibrium	withdrawn at intervals	
Analytes diffusion though carrot cylinders [23]	$E = \dfrac{C - C_L}{C_i - C_L} =$ $1 - \dfrac{4}{\pi^{\frac{1}{2}}}\left(\dfrac{D_e t}{a^2}\right)^{1/2} - \dfrac{D_e t}{a^2} - \dfrac{1}{3\pi^{\frac{1}{2}}}\left(\dfrac{D_e t}{a^2}\right)^{3/2}:$ Solution of Fick's 2nd law for solute loss or uptake for an infinite cylinder for short time periods. Authors then modify values to apply to finite cylinder	Carrot samples withdrawn for analysis at intervals	Dimensionless time values calculated for each data point using equation, linear regression of Fourier relationship yields D_e
Analytes diffusion though potato tissue [24]	$\dfrac{M_t}{M_i} = \dfrac{8}{\pi^2}\sum\limits_{n=0}^{\infty}\dfrac{1}{(2n+1)^2}exp\dfrac{-D_e(2n+1)^2}{a^2}\pi^2 t:$ Solution of Fick's 2nd law similar to flow through a membrane	Samples of solvent surrounding slices withdrawn and assayed at intervals, material balance used to calculate M_i and M_t	Equation model modified by omitting terms, $n''/>0$ and natural logarithms, by non-linear regression of $ln\, M_i/M_t$ against t
Analytes and pectic substances diffusion though apple tissue [25]	$E = \dfrac{C - C_L}{C_i - C_L} = \dfrac{M_t}{M_i} =$ $\dfrac{8}{\pi^2}\sum\limits_{n=0}^{\infty}\dfrac{1}{(2n+1)^2}exp\left[-(2n+1)^2\left(\dfrac{D_e t}{a^2}\right)\dfrac{\pi^2}{2}\right]:$ Solution of Fick's 2nd law for diffusion from an infinite slab developed by Newman [59]	Apple slices withdrawn for analysis at intervals	A series of dimensionless time values found for data using Newman's tables. D_e calculated by linear regression of Fourier relationship
Analytes diffusion though apple tissue [26]	$E = \dfrac{M_t}{M_i} = \dfrac{512}{\pi^6}exp\left[-\dfrac{\pi^2 D_e}{4(a^2+b^2+c^2)}\right]t:$ Equation model adopted considers diffusion in a slab in three planes.	Apple slices withdrawn for analysis at intervals	Modification of equation model by natural logarithms enables regression of lnE against t give slope with D_e term
Salt and acetic acid diffusion though herring [27]	$\dfrac{M_t}{M_\infty} = 1 - \sum\limits_{n=0}^{\infty}\dfrac{8}{(2n+1)^2\pi^2}exp\left[\dfrac{-D_e(2n+1)^2\pi^2 t}{4a^2}\right]:$ Solution of Fick's 2nd law for diffusion from/or uptake by infinite solution.	Fish withdrawn for analysis at intervals	Experimental data fitted to equation model by successive approximations
Analytes diffusion though potato tissue [28]	$\dfrac{M_t}{M_\infty} = \dfrac{V_L\, C_{Li} - V_L\, C_L}{V_L\, C_{Li} - V_L\, C_\infty} =$ $1 - \sum\limits_{n=1}^{\infty}\dfrac{2a(1+a)}{1+a+a^2 n^2}exp\dfrac{-D_e n^2 t}{a^2}:$	Initial solute content of potato strip and solute contents of solvent	Minimizing the residuals between experimental data and theoretical

Diffusion process	Diffusion	Experimental	Calculation
	equation	measurements	Diffusivity
	Solution of Fick's 2nd law for diffusion from/or uptake by an infinite slab for uptake from a solution of finite volume.	surrounding strip at intervals	values for M_t/M_∞ obtained from the appropriate equation models
	$\dfrac{M_t}{M_\infty} = 1 - \sum\limits_{n=0}^{\infty} \dfrac{8}{(2n+1)^2\pi^2} \exp\left[\dfrac{-D_e(2n+1)^2\pi^2 t}{4a^2}\right]$:		
	Solution of Fick's 2nd law for diffusion from/or uptake by infinite slab.		

Table 3. Diffusion phenomena.

The curve for the above solution tends to become linear at longer times (generally after $t > 0.5$ t_c), and ln (M_r/M_i) is given approximately by:

$$Ln(M_r / M_i) = -0.4977 - t / t_0 \tag{8}$$

Where t_c (min) is a characteristic time quantity, defined as:

$$t_c = r^2 / \pi^2 D_e \tag{9}$$

An alternative form of eq. 7, or so called a one-site kinetic desorption model, can be written for the ratio of mass of analyte removed after time t to the initial mass, M_i, as given by:

$$\frac{M_t}{M_i} = 1 - e^{-kt} \tag{10}$$

In which M_t is the mass of the analyte removed by the extraction fluid after time t (mg/g dry sample), M_i is the total initial mass of analyte in the matrix (mg/g dry sample), M_t/M_i is the fraction of the solute extracted after time t, and k is a first order rate constant describing the extraction (min^{-1}).

Matlab curve fitting solver was used to determine the desorption rate constant, k, from the data for all flow rates. The values for Z. *multiflora* SWE are show in Table 4 [11]. As mentioned, the kinetic desorption model does not include a factor describing extraction flow rate, k should be the same value for all flow rates if the model is said to fit the experimental data. However, this was not the case (Table 4, the average error 3%-17%). The kinetic desorption rate increased for the volumetric flow rate of 1 to 4 ml/min. This indicated that the kinetic desorption model may not be suitable for describing the data at different flow rates of Z. *multiflora*.

5.2.3. Two-site kinetic desorption model

Two-site kinetic model is a simple modification of the one-site kinetic desorption model that describes extraction which occurs from the "fast" and "slow" part [10]. In such case, a certain

| Flow rate/ | k/min^{-1} | |
ml·min^{-1}	Thymol	Carvacrol
1	0.0025	0.0028
2	0.0042	0.0039
4	0.0157	0.0157

Table 4. Values of k for one-site kinetic desorption model for different volumetric flow rates [7].

fraction (F) of the analyte desorbs at a fast rate defined by k_1, and the remaining fraction (1-F) desorbs at a slower rate defined by k_2. The model has the following form:

$$\frac{M_t}{M_i} = 1 - \left[Fe^{-k_1 t} \right] - \left[(1-F)e^{-k_2 t} \right]$$ (11)

The two site kinetic model does not include solvent volume, but relies solely on extraction time. Therefore, doubling the extractant flow rate should have little effect on the extraction efficiency when plotted as a function of time. On the contrary, the thermodynamic model is only dependent on the volume of extractant used. Therefore, the extraction rate can be varied by changing the flow rate. Hence, the mechanism of thermodynamic elution and diffusion kinetics can be compared simply by changing the flow rate in SWE. If the concentration of bioactive compounds in the extract increases proportionally with an increase in flow rate at given extraction time when the solute concentration is plotted versus extraction time, the extraction mechanism can be explained by the thermodynamic model. However, if an increase in flow rate has no significant effect on the extraction of the bioactive compounds, with the other extraction parameters being kept constant, the extraction mechanism can be modeled by the two site kinetic model [10, 31]. The mechanism of control and hence the model valid for SWE may be different depending on the raw material, the target analyte and extraction conditions.

For the two-site kinetic desorption model, the values of k_1 and k_2 were determined by fitting the experimental data with the two-site kinetic desorption models by minimizing the errors between the data and the model results. In the two-site model, the extraction rate should not be dependent on the flow rate. The k_1 and k_2 values for Z. *multiflora* SWE shown in Tables 5 and 6 demonstrated that the extraction rates were not completely independent of flow rate (the average error 11%-20%).

5.2.4. Thermodynamic partition with external mass transfer resistance model

This model describes extraction which is controlled by external mass transfer whose rate is described by resistance type model of the following form:

Flow rate/ml·min^{-1}	k_1/min^{-1}	k_2/min^{-1}	Mole fraction F
1	0.0088	0.0015	0.21
2	0.0152	0.0026	0.28
4	0.0770	0.0083	0.27

Table 5. k_1 and k_2 values of two-site kinetic desorption model for thymol at different flow rates [11].

Flow rate/ml·min^{-1}	k_1/min^{-1}	k_2/min^{-1}	Mole fraction F
1	0.0101	0.0017	0.21
2	0.0747	0.0088	0.42
4	0.0469	0.0082	0.27

Table 6. k_1 and k_2 values of two-site kinetic desorption model for carvacrol at different flow rates [11].

$$\frac{\partial C_s}{\partial t} = -k_e a_p \left[\left(C_s / K_D \right) - C \right] \tag{12}$$

in which C is the fluid phase concentration (mol/m^3), C_s is the solid phase concentration (mol/m^3), k_e is the external mass transfer coefficient (m/min) and a_p is specific surface area of particles (m^2/m^3) [32]. If the concentration of the solute in the bulk fluid is assumed small and the solute concentration in the liquid at the surface of solid matrix is described by partitioning equilibrium, K_D, the solution of eq. 11 for the solute concentration in the solid matrix, C_s, becomes:

$$C_s = C_0 - \exp(-k_e a_p t / K_D) \tag{13}$$

eq. 11 can be rewritten as the ratio of the mass of diffusing solute leaving the sample to the initial mass of solute in the sample, M_t/M_i, as given by the following equation.

$$M_t = 1 - M_i \exp(-k_e a_p t / K_D) \tag{14}$$

Because a_p is difficult to be measured accurately, a_p and k_e are usually determined together as $k_e a_p$, which is called overall volumetric mass transfer coefficient. The factors that influence the value of $k_e a_p$ include the water flow rate through the extractor and the size and shape of plant sample.

The values for the model parameters, K_D and $k_e a_p$ in eq. (9) determined by Matlab curve fitting solver from the experimental data obtained at 150°C are summarized for Z. *multiflora* SWE in

Tables 7 and 8 for different mass flow rates (Q, mg min^{-1}) [11]. Linear regresion of the plot between ln(k$_e$a$_p$) and lnQ gives the following correlation for k$_e$a$_p$ and Q:

$$\text{for thymol:} \qquad k_e a_p = 6.5748 \; Q_m^{0.2078} \tag{15}$$

$$\text{for carvacrol:} \qquad k_e a_p = 0.1605 \; Q_m^{0.6017} \tag{16}$$

Flow rate/ ml·min^{-1}	Mass flow rate Q_m/mg·min^{-1}	Parameter K_D	Parameter $k_e a_p$/min^{-1}
1	938	80	26.700
2	1876	80	32.7975
4	3752	2	1.300

Table 7. Parameters K$_D$ and k$_e$a$_p$ for external mass transfer model of SWE of thymol [11].

Flow rate/ ml·min^{-1}	Mass flow rate/mg·min^{-1}	Parameter K_D	Parameter $k_e a_p$/min^{-1}
1	938	70	8.92
2	1876	70	62.013
4	3752	2	20.54

Table 8. Parameters K$_D$ and k$_e$a$_p$ for external mass transfer model of SWE of carvacrol [11].

To quantitatively compare the extraction models, the mean percentage errors between the experimental data and the models were considered. Based on the result in fitting from experimental data, the K$_D$ model was generally suitable for the description of extraction over all the volumetric flow rates tested. On the other hand, one-site and two-site kinetic desorption models describe the extraction data reasonably at lower volumetric flow rates. Of all the models considered, however, the thermodynamic partition with external mass transfer model could best describe the experimental data.

5.2.5. Two-phase model

A mathematical model can be developed to predict optimal operating parameters for SWE in a packed-bed extractor. Three important steps consist of diffusion of solutes through particles, diffusion and convection of solutes through layer of stagnant fluid outside particles and elution of solutes by the flowing bulk of fluid are assumed. Unsteady state mass balance of the solute in solid and subcritical water phases led to two partial differential equations. The model can be solved numerically using a linear equilibrium relationship. The model parameters were mass transfer coefficient, axial dispersion coefficient, and intraparticle diffusivity. The last parameter was selected to be the model tuning parameter. The two other parameters were predicted applying existing experimental correlations.

5.2.5.1. Model description

The more precise method is based on differential mass balances along the extraction bed. A two-phase model comprising solid and subcritical water phases can be used. Extraction vessel is considered to be a cylinder filled by mono-sized spherical solid particles. The overall scheme of system was like Fig. 7.

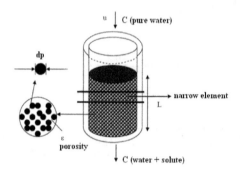

Figure 7. Characteristic dimensions and geometry of the packed bed SWE vessel.

The major assumptions used to describe the SWE process for deriving the essential oils extraction model were:

1. Packed bed extractor was isothermal and isobaric,

2. The physical properties of subcritical water were constant,

3. The hydrodynamics of a fluid bed was described by the dispersed plug-flow model,

4. The radial concentration gradient in the bulk fluid phase was assumed to be negligible,

5. The volume fraction of bed was not influenced by the weight loss of plant during the extraction,

6. The essential oil was assumed as a single component.

Under these assumptions, the differential mass balance equation for any component in the particle and bulk liquid phase and associated initial and boundary conditions can be written as following dimensionless forms:

- Solid phase:

$$\frac{\partial C_p}{\partial \tau} = \frac{2}{Pe_p} \frac{L}{R} \left(\frac{\partial^2 C_p}{\partial Y^2} + \frac{2}{Y} \frac{\partial C_p}{\partial Y} \right) \tag{17}$$

$$\tau = 0 \qquad C_p = C_{p0} \tag{18}$$

$$Y = 0 \qquad \frac{\partial C_p}{\partial Y} = 0 \tag{19}$$

$$Y = 1 \quad \frac{\partial C_p}{\partial Y} = Sh \left(C_f - C_{fp} \right) \tag{20}$$

- Subcritical water phase

$$\frac{\partial C_f}{\partial \tau} = \frac{1}{Pe_b} \frac{\partial^2 C_f}{\partial X^2} - \frac{\partial C_f}{\partial X} - \frac{6(1-\varepsilon)L}{\varepsilon R} \frac{Bi}{Pe_p} \left(C_f - C_{fp} \right) \tag{21}$$

$$\tau = 0 \qquad C_f = 0 \tag{22}$$

$$X = 0 \qquad C_f - \frac{1}{Pe_b} \frac{\partial C_f}{\partial X} = 0 \tag{23}$$

$$X = 1 \quad \frac{\partial C_f}{\partial X} = 0 \tag{24}$$

Eqs. (17) and (21) may be solved using a linear equilibrium relationship between concentrations in the solid phase and SW phase at the interface, as follows [33]:

$$C_{fp} = k_p C_p |_{Y=1} \tag{25}$$

Where C_{fp} is the solute concentration in the fluid phase at the particle surface, $C_p|_{Y=1}$ is the solute concentration in the solid phase at equilibrium with the fluid phase and k_p is the volumetric partition coefficient of the solute between the solid and the fluid phase. Therefore, there are three eqs. (17), (21) and (25) which can be solved, simultaneously, for three unknowns C_p, C_f and C_{fp}.

The finite difference equations are a set of simultaneous linear algebraic equations must be solved for implicit method to obtain the concentration distribution at any time. In both cases, tridiagonal systems arise which are conveniently solved at each time step by the Thomas algorithm [34]. A computer code was written using MATLAB simulation software.

5.2.5.2. Parameter identification and correlations

The possible control of mass transfer was assayed by estimating the diffusion coefficient in the liquid. We can use the correlation proposed by Wilke and Chang (1955) to estimate this coefficient [35]:

$$D_{AB} = \frac{7.4 * 10^{-8}}{\mu V_A^{0.6}} \left(\varphi M_2 \right)^{0.5} T \tag{26}$$

Where φ is 2.26 for water and 1.5 for ethanol and V_A was estimated by the Tyn Calus equation:

$$V_A = 0.285 \, (V_c)^{1.048} \tag{27}$$

Where V_c is the estimated by the method of Joback and Reid (1987), with the aid of Molecular Modeling Plus software (Norgwyn Montgomery Software, USA) [36]. The axial dispersion coefficient in the supercritical phase was approximated as follows [37] and it may be used for the subcritical phase:

$$D_L = \frac{u \, d_p}{\varepsilon \, Pe_{pd}} \tag{28}$$

Where the average void volume fraction of the fixed bed was $\varepsilon = 0.4$ and:

$$Pe_{pd} = 1.634 \, Re^{0.265} \, Sc^{-0.919} \tag{29}$$

Intraparticle diffusivity of the essential oils in the solid phase, D_m, was selected to be the tuning parameter of the model. Mass transfer between liquids and beds of spheres, k_f, can be represented by Wilson and Geankoplis in two cases [38]:

$$\varepsilon \, j_D = 0.0016 < Re < 55, \quad 165 < Sc < 70600 \tag{30}$$

$$\varepsilon \, j_D = \frac{0.25}{Re^{0.31}} 55 < Re < 1500, \quad 165 < Sc < 10690 \tag{31}$$

$$k_f = \frac{j_D \, U_o}{Sc^{2/3}} \tag{32}$$

Density of water at high pressures and temperatures from 273 to 473 K was assumed to be the density of saturated water (kg). It was calculated as follows [39]:

$$\rho = 858.03 + 1.2128 \, T - 0.0025 \, T^2 \tag{33}$$

Where ϱ is density (kg/m³) and T is temperature (K). The water viscosity at temperatures from 300 to 450 K was calculated by the following equation:

$$\mu = \exp\left(-10.2 + \frac{280970}{T^2}\right) \tag{34}$$

Where μ is the viscosity (Pa.s) and T is temperature (K). Viscosity of water was supposed to be independent of pressure. The model was verified successfully using the SWE data for Z.multiflora leaves at 20 bar and 150°C. The optimum value of 2×10^{-12} m²/s was obtained for the intraparticle diffusivity [12].

5.2.6. CFD model

There is no scientific literature about the application of CFD modeling approach in the SWE processes. In our previous work, we have tried to do CFD modeling of essential oils of Z. *multiflora* leaves [13].

In CFD modeling in packed bed reactor, there are two cases. In first one, when the reactor to particle diameter ratio is less than10, it is needed to build the exact geometry and the location of the particles and solving the governing equations in meshes and can see the field of flow between particles inside reactor. But in another one, when the reactor to particle diameter ratio is higher than10, this system is defined as porous media. For modeling of these reactors, the Navier-stocks with one additional term, which is contained viscous force and inertia loss, in porous media is solved.

In this work, because of the reactor to particle diameter ratio is higher than10; the system can define as porous media. The momentum equation is defined as:

$$\frac{\partial}{\partial t}\left(\varepsilon\rho\vec{v}\right)+\nabla.\left(\varepsilon\rho\vec{V}\vec{V}\right)=-\varepsilon\nabla P+\nabla.\left(\varepsilon\tau\right)=\left(-\frac{\mu}{\beta}\vec{V}+\frac{1}{2}C\rho\left|\vec{V}\right|\vec{V}\right) \tag{35}$$

Where $\frac{\partial}{\partial t}(\varepsilon\rho\vec{v})+\nabla.(\varepsilon\rho\vec{V}\vec{V})=-\varepsilon\nabla P+\nabla.(\varepsilon\tau)=\left(-\frac{\mu}{\beta}\vec{V}+\frac{1}{2}C\rho\mid\vec{V}\mid\vec{V}\right)$ and C and β coefficient can define as:

$$\beta=\frac{d_P^2\varepsilon^3}{150\left(1-\varepsilon\right)^2} \tag{36}$$

$$C=\frac{3.5\left(1-\varepsilon\right)}{d_P\varepsilon^3} \tag{37}$$

In this work the porosity of bed is 0.4, so we have following value for C and β coefficients.

$$C=54687.5 \quad 1/m \tag{38}$$

$$\beta=0.23*10^{-10}1/m^2 \tag{39}$$

The energy equation is defined as:

$$\frac{\partial}{\partial t}\left(\varepsilon\rho_f E_f+\left(1-\varepsilon\right)\rho_s E_s\right)+\nabla.\left(\vec{V}\left(\rho_f E_f+P\right)\right)=\nabla.\left[\hat{K}\nabla T+\left(\tau.\vec{V}\right)\right]+S \tag{40}$$

Where the Effective thermal diffusion coefficient is defined as:

$$\hat{K} = \varepsilon \hat{K}_f + (1-\varepsilon)\hat{K}_s \qquad (41)$$

The spices transfer equation is defined as:

$$\frac{\partial c_i}{\partial t} + \frac{\partial}{\partial x}\left(-D_x\frac{\partial c_i}{\partial x}\right) + \frac{\partial}{\partial x}\left(-D_x\frac{\partial c_i}{\partial x}\right) = \frac{D_y}{y}\frac{\partial c_i}{\partial y} - \frac{u}{\varepsilon}\frac{\partial c_i}{\partial x} - \frac{u}{\varepsilon}\frac{\partial c_i}{\partial y} + S \qquad (42)$$

S is source term. Comparing the default of spices transfer equation with spices transfer equation of this system can define the

$$\frac{\partial c_i}{\partial t} + \frac{\partial}{\partial x}\left(-D_x\frac{\partial c_i}{\partial x}\right) + \frac{\partial}{\partial x}\left(-D_x\frac{\partial c_i}{\partial x}\right) = \frac{D_y}{y}\frac{\partial c_i}{\partial y} - \frac{u}{\varepsilon}\frac{\partial c_i}{\partial x} - \frac{u}{\varepsilon}\frac{\partial c_i}{\partial y} + S$$ as the source term of this work.

C_{fs} was the concentration of Thymol & Carvacrol in the fluid face at the particle surface and was defined as:

$$c_{fs} = k_p c_{ss} \qquad (43)$$

and

$$c_{ss} = 0.8\, c_0 \exp(-0.0005\,t) \qquad (44)$$

In such works, the Reynolds number is defined as:

$$Re_d = \frac{\rho \bar{U}_p \bar{d}_p}{\mu} \qquad (45)$$

$Re_d = \frac{\rho \bar{U}_p \bar{d}_p}{\mu}$ is mean characteristic length of porous and $\bar{d}_p$ is mean velocity based on porous. Based on this Reynolds number, it can be seen four regimes:

$Re_d<1$ Darcy regime or creeping flow

$1<Re_d<150$ inertia regime

$150<Re_d<300$ unsteady laminar flow regime

$Re_d>300$ unsteady Turbulent flow regime

The superficial velocity in this packed bed is defined as dividing flow rate by cross section.

Dividing superficial velocity by porosity, real velocity can calculate.

$$U = \frac{Q}{A} \tag{46}$$

$$\bar{U}_p = \frac{U}{\varepsilon} \tag{47}$$

and for calculating D_p

$$\bar{D}_p = \frac{6}{A_0} \tag{48}$$

In this work the mean particle diameter was 0.6 mm.

$$A_0 = \frac{A}{V} = \frac{4\pi r^2}{\frac{4}{3}\pi r^3} \tag{49}$$

In this work the Reynolds number is 0.2, so there is laminar flow in this system.

The material of system was mixture of Thymol & Carvacrol and water. The inlet flow was contained water with mole fraction 1 and the outer flow was consisted of water and Thymol & Carvacrol [13]. The packed bed had 103 mm height and the diameter of bed was 16 mm. The uniform mesh was used for this domain.

The governing equations are solved by a finite volume method. At main grid points placed in the center of the control volume, volume fraction, density and spices fraction are stored. The conservation equations are integrated in space and time. This integration is performed using first order upwind differencing in space and fully implicit in time. For a first-order upwind solution, the value at the center of a cell is assumed to be an average throughout the cell. The SIMPLE algorithm is used to relate the velocity and pressure equations.

The simulation geometry is shown in Fig.8. The 2D calculation domain is divided into 22*146 grid nodes, in the radial and axial directions, respectively. The grid and mesh size are chosen to be uniform in the two directions. The inlet of system was pure water and the outlet of system was extracted Thymol & Carvacrol.

The extraction values versus time for 150°C were shown in Fig. 9. It can be seen, the extraction values were increased as exponentially. After 60 min, the change of extraction values with time was very little.

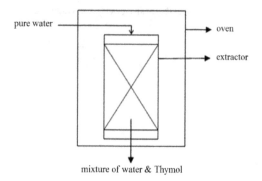

Figure 8. The Schematic of extractor [13].

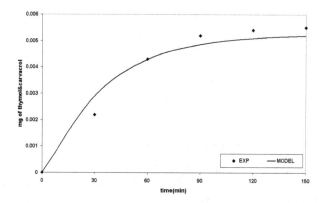

Figure 9. Extraction values versus time for 150°C [13].

In order to investigate the applicability of the CFD model, the theoretical results are compared with experimental measurements obtained at optimum conditions (20 bars, 150°C, and 2 ml/ min). Fig. 9 shows that the modeling extraction values profile is increasing rapidly in the period of 0-60 min and thereafter in the second region (60-120 min) the slope reduces until reaches a constant trend in the third period of 120-150 min. In the first region, because of high Thymol & Carvacrol concentrations in the Z. *multiflora* leaves and therefore, high mass transfer driving force, high desorption rate of Thymol & Carvacrol from solid matrix occurs.

6. Conclusion

It was tried to give overall view about subcritical water extraction. Effective parameters, mechanism and modeling of extraction were surveyed. Overall by considering mean average errors of models, a mathematical model base on the combination of partition coefficient (KD) and external mass transfer gave a good description of subcritical water extraction of Z. *multiflora*, while the kinetic model reasonably described the extraction behavior at lower flow rates [11].

On the other side, model was developed by introducing differential mass balances using two phase model, and applying a linear equilibrium relationship. Because of considering the effect of variation of the concentration profile in the SW phase, it seems that the proposed model is more significant from the physical point of view.

The Last model was CFD modeling of extraction from Z. *multiflora* leaves using subcritical water. It was concluded that CFD is poised to remain at the forefront of cutting edge research in the sciences of fluid dynamics and mass transfer. Also, the emergence of CFD as a practical tool in modern engineering practice is steadily attracting much interest and appeal. The results of CFD model have been agreed well with experimental data. As shown, along of extractor, Thymol was extracted and was in outflow.

Notations

specific surface, m^2/m^3 ($= \frac{3}{R_p}$)	a or a_p
Biot number ($= \frac{k_f R_p}{D_m}$)	Bi
solute concentration in the solid phase, $kmol/m^3$	C
solute concentration in the SW phase, $kmol/m^3$	C_f
solute concentration in the fluid phase at the particle surface, $kmol/m^3$	C_{fp}
solute concentration in the solid phase, $kmol/m^3$ ($\frac{C_p}{C_{po}}$)	C_p
initial solute concentration in the solid phase, $kmol/m^3$	C_{po}
diffusivity of solute (A) in liquid (B), m^2/s	D_{AB}
effective diffusion coefficient (m^2/s)	D_e
axial dispersion coefficient, m^2/s	D_L
diffusivity in the solid, m^2/s	D_m
particle diameter, m	d_p
exponential function	e
a certain fraction of the analyte desorbs at a fast rate by k_1	F
remaining fraction desorbs at a slower rate by k_2	$(1-F_1)$
thermodynamics partitioning coefficient	K_D
external mass transfer coefficient (m/min)	k_e

mass transfer between liquids and beds of spheres	k_f
volumetric partition coefficient of the solute between the solid and the fluid phase	k_p
cumulative mass of analyte extracted after certain amount of volume V_a (mg/g dry sample)	M_a
cumulative mass of the analyte extracted after certain amount of volume V_b (mg/g dry sample)	M_b
total initial mass of analyte in the matrix (mg/g dry sample)	M_i
cumulative fraction of the analyte extracted by the fluid of the volume V_b and V_a (ml)	M_b/M_i and M_a/M_i
total amount of solute (mg/g) removed from particle after time t	M_t
maximum amount (mg/g) of solute extracted after infinite time	M_∞
ratio of total migration to the maximum migration concentration	M_t/M_∞
mass of the extracted sample (mg dry sample)	m
Peclet number of the bed ($= \frac{u_0 L}{D_L \varepsilon}$)	Pe_b
Peclet number of the solid ($= \frac{u\, d_p}{D_m \varepsilon}$)	Pe_p
average radius of an extractable particle	R
Reynolds number ($= \frac{2 R u \rho}{\mu}$)	Re
Schmidt number ($= \frac{\mu \rho}{D_{AB}}$)	Sc
Sherwood number ($= \frac{2 kf R}{D_{eff}}$)	Sh
Stanton number ($\frac{L (1 - \varepsilon) k_f\, a}{u}$)	St
Temperature, K	T
superficial SW fluid velocity, m/s	u
molar volume of the solute at its normal boiling point, cm³/mol	V_A
critical volume, cm³/mol	V_c
dimensionless axial coordinate along the bed, z/L	X
dimensionless radius ($= \frac{r}{R}$)	Y

Greek symbols

ε	void volume fraction
μ	viscosity, Pa.s
ρ	density, kg/m³
φ	association factor for the solvent

τ	dimensionless time ($= \frac{u\,t}{L\,\varepsilon}$)

Acknowledgements

This research was supported by Semnan University. The authors would like to thanks the Office of Brilliant Talents at the Semnan University for financial support.

Author details

A. Haghighi Asl and M. Khajenoori

School of Chemical Gas and Petroleum Engineering, Semnan University, Semnan, I.R., Iran

References

[1] C.W. Huie, "A review of modern sample-preparation techniques for the extraction and analysis of medicinal plants," Anal. Bioanal. Chem. 373 (2002) 23-30.

[2] B. Zygmunt, J. Namiesnik, "Preparation of samples of plant material for chromatographic analysis," J. Chromatogr. Sci. 41 (2003) 109-116.

[3] E.S. Ong, "Extraction methods and chemical standardization of botanicals and herbal preparations," J. Chromatogr. B 812 (2004) 23-33.

[4] King, et al., US Patent 7,208,181, B1, 2007.

[5] R. M., Smith, "Superheated water: The ultimate green solvent for separation science", Anal. Bioanal. Chem., 385(3), 419 (2006).

[6] D. J., Miller, S. B., Hawthorne, "Solubility of liquid organics of environmental interest in subcritical (hot/liquid) water from 298 K to 473 K", J. Chem. Eng. Data, 45, 78 (2000).

[7] M. Khajenoori, A. Haghighi Asl, and F. Hormozi, M. H. Eikani and H. Noori. "Subcritical Water Extraction of *Zataria Multiflora Boiss.*", *Journal of Food Process Engineering,* 32, (2009) 804–816.

[8] M.H., Eikani, F., Golmohammad and S., Roshanzamir, "Subcritical water extraction of essential oils from coriander seeds (*Coriandrum sativum Mill.*)", J. Food Eng. 80 (2) (2007a) 735-740.

[9] M.H., Eikani, F., Golmohammad, S., Roshanzamir and M. Mirza, "Extraction of vola-
     tile oil from cumin (*Cuminum cyminum L.*) with superheated water", J. Food Process
     Eng., 30 (2) (2007b) 255-266.

[10] A., Kubatova, B., Jansen, J.F., Vaudoisot, S. B., Hawthorne, "Thermodynamic and ki-
     netic models for the extraction of essential oil from savory and polycyclic aromatic
     hydrocarbons from soil with hot (subcritical) water and supercritical CO_2", J. Chro-
     matography A., 975(1), (2002) 175-188.

[11] M. Khajenoori, A. Haghighi Asl, and F. Hormozi, "Proposed Models for Subcritical
     Water Extraction of Essential Oils", Chinese Journal of Chemical Engineering, Vol.
     17, No. 3, (2009) 359-365.

[12] M. Khajenoori, A. Haghighi Asl, and M. H. Eikani "Modeling of Superheated Water
     Extraction of Essential Oils". Submitted in "13th Iranian National Chemical Engi-
     neering Congress & 1st International Regional Chemical and Petroleum Engineering
     Kermanshah, Iran, 25-28 October, 2010".

[13] M. Khajenoori, E. Omidbakhsh, F. Hormozi, and A. Haghighi Asl, "CFD modeling of
     subcritical water extraction". the 6th International chemical Engineering Congress
     (IChEC), Kish Island, Iran, 16-20 November 2009.

[14] J., Crank,. "The mathematics of Diffusion". Oxford, England: Clarendon Press. (1975)
     pp. 150-175.

[15] V., Gekas, "Transport phenomena of foods and biological material". Boca Raton, FL.
     CRC Press., (1992) pp. 156-178.

[16] E. L., Cussler, "Diffusion: Mass Transfer in Fluid Systems". Cambridge University
     Press. Cambridge, UK., (1984) pp. 146-177.

[17] C., Mantell, M., Rodriguez, and E. Martinez de la Ossa, "Semi-batch extraction of an-
     thocyanins from red grape pomace in packed beds: experimental results and process
     modeling", Chem. Eng. Sci., 57 (2002). 3831-3838.

[18] Dibert, K., Cros, E., and Andrieu, J. Solvent extraction of oil and chlorogenic acid
     from green coffee. Part II. Kinetic data. J. Food Eng., 10, (1989) 199-214.

[19] Tutuncu, M. A., and labuza, T.P.. Effect of geometry on the effective moisture trans-
     fer diffusion coefficient. J. Food Eng. 30: (1996) 433-447.

[20] J.A. Bressan, P.A. Carroad, R.L. Merson, W.L. Dunkley, Temperature dependence of
     effective diffusion coefficient for total solids during washing of cheese curd. J. Food
     Sci, 46 (1958) 9.

[21] E.S.A., Biekman, C., Van Dijk, "Measurement of the apparent diffusion coefficient of
     proteins in potato tissue", Presented at DLOA grotechnological Research Institute,
     The Netherlands, (1992).

[22] C.R., Binkley, RC., Wiley, "Chemical and physical treatment effects on solid–liquid extraction of apple tissue", J. Food Sci., 46 (1981) 729-732.

[23] NS., Kincal, F., Kaymak, "Modeling dry matter losses from carrots during blanching", J. Food Process Eng., 9 (1987) 201–211.

[24] I., Lamberg, 'Transport phenomena in potato tissue". Ph.D. thesis, University of Lund, Lund, Sweden; (1990).

[25] G.C., Leach, DL., Pyle, K., Niranjan, 'Effective diffusivity of total solids and pectic substances from apple tissue", Int. J. Food Sci. Technol., 29 (1995) 687–897.

[26] A., Lenart, PP., Lewicki, J., Dziuda, "Changes in the diffusional properties of apple tissue during technological processing". In: Spiess WEL, Schubert H., editors. Engineering and food, vol.1: Physical properties and process control. London: Elsevier Applied Science; (1989) pp. 531-540.

[27] G., Rodger, R., Hastings, C., Cryne, J., Bailey, "Diffusion properties of salt and acetic acid into herring and their subsequent effect on the muscle tissue". J. Food Sci., 49 (1984) 714-732.

[28] P., Tomasula, MF., Kozempel, "Diffusion coefficients of glucose, potassium, and magnesium in Maine Russet Burbank and Maine Katahdin potatoes from 45 to 90°C", J. Food Sci. 54 (4) (1989) 985–999.

[29] H.G., Schwartzberg, R.Y., Chao, "Solute diffusivities in leaching process", Food Technol., 36, (1982) 73-86.

[30] D.D., Gertenbach, "Solid-liquid extraction technologies for manufacturing nutraceuticals", Shi, J., Mazza, G., Maguer, M.L., eds., Functional Foods: Biochemical and Processing Aspects (Vol.2), CRC Press, Boca Raton, Flordia (2002).

[31] J.E., Cacace, G., Mazza, "Pressurized low polarity water extraction of lignans from whole flaxseed", J. Food. Eng., 77 (2006) 1087-1095.

[32] Th., Anekpankul, M., Goto, M., Sasaki, P., Pavasant, A., Shotipruk, "Extraction of anti-cancer damnacathal from roots of *Morinda citrifolia* by subcritical water", Sep. Purif. Technol., 55 (2007) 343-349.

[33] E., Reverchon, "Mathematical modeling of supercritical extraction of sage oil," A.I.Ch.E. Journal, 42 (1996).1765-1771.

[34] Y. Jaluria, "Computer Methods for Engineering". Allyn and Bacon, Newton, MA, U.S.A. (1988).

[35] C. R. Wilke, P. Chang, "Correlation of diffusion coefficients in dilute Solutions", AICHE Journal, 1 (1955) 264-270.

[36] K. Joback, R. Reid, "Estimation of pure component properties from group contributions", Chemical Eng. Commun., 57 (1987) 233-243.

[37] C. S. Tan and D. C. Liou, "Axial dispersion of Supercritical carbon dioxide in packed beds," Ind. Eng. Chem. Res., 28 (1989) 1246-1250.

[38] J. R. Welty, C. E. Wicks, R. E. Wilson, "Fundamentals of Momentum, Heat, and Mass Transfer", 3rd ed., John Wiley & Sons, New York, (1984).

[39] R. Perry, D. W. Green, J. O. Maloney, "Perry Chemical Engineers Handbook," 5rd ed. McGraw-Hill, New York (1984).

Chapter 9

Mass Transfer:
Impact of Intrinsic Kinetics on the Environment

Engr Owabor

Additional information is available at the end of the chapter

1. Introduction

Emerging trends in environmental engineering and biosystem analysis indicate that a clean, safe and healthy ecosystem is still possible regardless of the increased quest for technology and industrialization by a dynamic society where the fuel requirements of the energy sector and the chemical and petrochemical needs of the chemical and allied industries are greatly dependent on the petroleum industry in which the downstream sector is a key player. These needs range from the feedstock requirements of the chemical industries, polymeric industrial sector, the solvent sector, the cosmetics industries etc. The concern over industrial and technological expansion, energy utilization, waste generation from domestic and industrial sources viz a viz their attendant negative effects on the environment cannot be overemphasized.

This chapter is intended to expose the reader to the role of mass transfer on the topical issue of environmental remediation. This is against the backdrop of the interrelationship between mass transfer and the major processes such as biological, chemical and phyto-oxidation acting during the removal of toxic substances from the environment and the dependence of these processes on the availability of these substances which over time compromise the integrity of the environment. Research reports advanced in this chapter is aimed at highlighting the crucial need to very seriously exploit the principles of mass transfer in addressing the critical issue of environmental degradation. The underlying mechanisms involved in the removal process and the systematic protocol developed for accelerated clean up are incorporated.

Basically, mass transfer is one of the important three transport operations commonly encountered and employed in chemical transformations. It has been described fundamentally, as the movement of the components of a mixture from one point to another as a result of observed differences in concentration [1]. The material transfer is usually in the direction of decreasing concentration gradient.

The concept of mass transfer is very important in the fields of engineering (particularly Chemical Engineering) and sciences in general where it is applied for physical processes. From the standpoint of chemical engineering, mass transfer is of prime significance because most of the unit operations involving separation of mixtures into their component parts employ mass transfer operation. The relevance of mass transfer operations is continuously a subject of discourse as specifically, there is hardly any chemical process which does not require a preliminary purification of raw materials or final separation of desired products from by-products. In essence it can be described as the process that determines the rate at which separation will occur. Issues relating the removal of certain materials from fluid streams typically by adsorption are important industrial applications of the mass transfer operation.

The challenges of maintaining a clean environment stem primarily from the increasing pollution of waters and land by heavy metals that are difficult to decompose biologically and organic compounds which are persistent, recalcitrant and ubiquitous. This is because these substances resist the self-purification capabilities of the rivers as well as decomposition in conventional wastewater treatment plant. Consequently, conventional mechanical-biological purification no longer suffices. It thus becomes imperative to supplement an additional stage of processing and this improvement has been augmented by adsorption; an innovative treatment method based on the principle of diffusion (a mechanism of mass transfer). Litera-ture reports have shown that the mechanism of the process of material transfer is by diffusion and this is dependent to a large extent on the physical characteristics and attributes of the mass been considered for transfer. The transfer process which is dependent on concentration gradient is basically of two types; molecular diffusion and convective transfer. While the former is known to be involved with movement of a material between two points i.e from a high to low area of concentration, the latter relates to the flow of liquids around a solid surface in forced convection motion. Interestingly, the two transfer types are governed by Fick's first law of diffusion under the condition of unchanging concentration gradient with the passage of time. However, Fick's second law becomes applicable under unsteady-state diffusion. The principles of Fick's laws which suggest that the rate of diffusion is a linear function of the solute concentration gradient has continuously been affirmed by research results involving material transport. In general, mass transfer cannot really be separated from adsorption and desorption processes as the transport of materials from one phase to another is principally by the mech-anism of diffusion. Sorption is the primary process in the evaluation of availability while mass transfer is the mechanism of movement from the fluid phase to the surface of the soil particle. Various isotherm correlations have been used to describe the surface adsorption of the solutes and their subsequent desorption from the soil surface [1]. The sorption of contaminant tends to separate the direct contact between microorganisms, adsorbents and contaminants, which is necessary for biodegradation detoxification to occur. The practical effect of the adsorption and desorption rate, is that it controls the overall reaction rate. The transport (mobility) of contaminant solute therefore is significantly dependent on two possible scenarios: fast sorption/desorption and slow sorption/ desorption.

Development of surface modified activated carbon from low cost and readily available natural materials and agricultural wastes have been advanced for the generation of activated carbon with far superior adsorption capacity for the treatment of contaminated water. The removal of pollutants from aqueous waste stream by adsorption in fixed beds has therefore progressively become an important treatment process and a subject of many excellent research works.

Laboratory and field studies results abound to demonstrate the applications of mass transfer in the management of the environment in terms of pollution monitoring and control. The intrinsic kinetics of the mass transfer process has through systematic and sustained researches become a relevant area of interest which provides information on availability, mobility and toxicity as a function of the measured concentration and the mechanism of sequestration of the solutes in the containing system.

2. Applications of mass transfer in water treatment technology

The demand for clean water for fishing activities, recreation etc and its non-ready availability has caused considerable attention to be focused towards recovery and re-use of waste waters. In water treatment procedures where the process of adsorption has been effectively engaged, the use of adsorbent materials has served as the solid surface onto which molecules of the adsorbates such as organic compounds, heavy metals, attach themselves. The removal of contaminant solutes from the environment has not been effectively accomplished by traditional methods. Hence it has now become universally recognized that adsorption technology provides a feasible and effective method for the removal of pollutants from polluted water resources and waste waters. Activated carbons are the most commonly used adsorbent in the adsorption process due to their high adsorption capacity, high surface area and high degree of surface reactivity. Adsorbents have been sourced variously from refinery residue, coal but mostly from agricultural waste materials such as coconut shell, soya bean, cotton seed, walnut and rice hulls, melon husk, orange peels etc.

The effectiveness of low density adsorbents dried (ground ripe and unripe orange peels) in the removal organics typified using naphthalene and pyrene from an aqueous stream has been extensively studied [2]. The influence of the variation in process conditions such as concentration, adsorbent dosage, agitation time and particle size, pH, and initial solute concentration were analyzed to highlight the contributions of mass transfer in the treatment of contaminated water. Under the operating conditions specified above, laboratory and experimental results showed that the removal efficiency defined by adsorption capacity increased with an increase in adsorbent dosage, contact time and initial solute concentration but decreased with an increase in the particle size. This experimental result clearly demonstrates that the transfer of any organic compound under consideration from one point to another is a linear function of the operating variables which positively improve the reliability of the process.

Following the same adsorption principle for mass transfer, the removal of mono and polycyclic aromatic hydrocarbons (PAHs) from bulk fluid stream onto adsorbents from periwinkle and coconut shell (PSC) also provides an alternative technology for effective mass transfer process in the reduction of pollutant level of wastewater. Naphthalene, phenanthrene and anthracene were used as the representative PAHs and the effluent stream was simulated refinery wastewater.

The result for the characterization of the refinery effluent before and after treatment with periwinkle shell activated carbon showed that the total suspended solids (TSS) reduced from an initial value of 53 mg/l to 2 mg/l after treatment.

Further insights into the role of adsorption on the contaminated water has revealed that the biochemical and chemical oxygen demand of the wastewater reduced drastically from an initial value to specifications recommended by the environmental protection agency (EPA) using activated carbon from materials such as periwinkle shell [3]. As contact time increased, percentage removal of the organic and inorganic contaminants present in the wastewater also increased. From the bench scale results [3], the total dissolved solids and total hydrocarbon content of the wastewater treated with activated periwinkle shell carbon reduced significantly. A maximum percentage removal of 78.5% was achieved. The clarity of the effluent water was observed to be greatly improved after the first hour of treatment where the turbidity dropped from 66 NTU to 6 NTU. Again the percentage turbidity increased with contact time until 100% removal was achieved following complete clarity.

The result showed comparatively, that the treated effluent water had maximum percentage reduction of approximately 83.26%, 91.19% (BOD, COD) for periwinkle shell carbon (PSC) and commercial activated carbon (CAC) respectively. The periwinkle shell carbon displayed the capacity to reduce the turbidity of the water contaminated with organic compounds via transfer of the contaminant solute from the bulk wastewater where their concentration is high to the external surface of the activated periwinkle shell carbon acting as the adsorbent.

In the development of alternative fixed filter bed for water treatment, the application of the mass transfer process by adsorption has also been shown to be feasible by examining the influence of modified clay on the rates of adsorption. The modifiers used were inorganic and organic acids, bases and salts. The results obtained showed that equilibrium adsorption of naphthalene i.e the transfer of naphthalene molecules from the bulk solution was attained at a faster rate using modified clay when compared with the unmodified clay. Amongst the modified clay used for the study, acids were found to be most suitable for the modification purpose [4].

This is attributable to the fact that the solutes in the acid medium had higher diffusive mobility defined by their dispersion coefficient/diffusivity as shown in Table 1 and higher influence on the physicochemical properties (porosity, surface area, bulk and particle density).

Overall, results from this study affirm that sediment modification directly affects the availability of the contaminant chemicals.

3. Applications in soil remediation

In soil remediation, the availability of the contaminant is of prime significance and it is commonly approached from the premise that chemicals are immediately accessible for

Modifying agents	D_{AB} (cm^2/s)
HCl	4.88×10^{-6}
H$_2$SO$_4$	4.5×10^{-6}
HNO$_3$	4.33×10^{-6}
H$_3$PO$_4$	2.97×10^{-6}
KOH	5.15×10^{-7}
NaOH	4.36×10^{-7}

*Adapted from Oladele, A.O. (2012). M.Eng. Thesis. University of Benin, Nigeria.

Table 1. Parameter estimation for naphthalene diffusivity in modified-clay sediment*

microbial uptake only when in aqueous solution. The rationale behind this premise is that the small pore spaces internal to aggregates of soil and sediment particles exclude microbes, such that compounds that are dissolved or sorbed within these immobile-water domains must first be transported to the external aqueous phase (i.e., bulk aqueous phase) before they can be metabolized [5]. This premise is supported by several laboratory studies that have found biodegradation to occur only, or predominantly, in the bulk aqueous phase [6], [7]. A variety of factors, including physical characteristics of the sorbent (e.g., particle shapes, sizes, and internal porosities), chemical properties of the sorbates and sorbents, and biological factors (e.g., microbial abundance and affinity for the contaminant) influence availability.

In a study conducted [8], the experimental data from a soil microcosm was analyzed using the method of temporal moments (MOM) which interprets solute transport with linear equilibrium sorption and first order degradation and the analytical solutions of a transport model CXTFIT version 2.0. This was with a view to estimating the transport parameters (pore-water velocity, V and dispersion coefficient, D) using non-reactive solute and the degradation parameters (retardation factor R and first order degradation rate λ) of the contaminant PAHs.

A dimensionless parameter called the retardation factor was used to represent bioavailability (i.e the accessibility of a chemical for assimilation and possible toxicity). This parameter increased with increasing solute hydrophobicity.

The result from the study is summarized in Table 2 and it showed that naphthalene had the lowest retardation factor with a corresponding higher degradation rate constant. The observed trend is similar and comparable to previous estimates from the Michaelis-Menten kinetics. The finding from this study can be attributed to the aqueous solubility, diffusivity and mobility of each contaminant solute.

PAHs	R			λ (per day)		
	MOM	CXTFIT	ε	MOM	CXTFIT	E
Naphthalene	25.77	20.23	0.21	3.54	4.22	-0.19
Anthracene	41.62	28.43	0.32	1.21	2.05	-0.69
Pyrene	35.66	25.89	0.27	2.25	3.26	-0.45

*Adapted from Owabor, C.N. (2007) PhD Thesis, University of Lagos, Nigeria.

Table 2. Comparison of the Degradation and Transport parameters for the Contaminant PAHs*

In general, the result is significant as it affirms that the biodegradation of the contaminant organic compound or specifically PAHs is a function of their bioavailability.

Studies have led researchers to conclude that some microorganisms are capable of degrading compounds directly from the sorbed phase. Also, for many studies where solid-phase degradation was reported, biodegradation rates were nonetheless observed to decrease with contaminant soil-water contact time. Access to contaminants is increasingly inhibited as solutes migrate deeper into submicron pores of impermeable sorption domains in soil/ sediment-water environments. From the foregoing, if solutes are degraded in the mobile aqueous phase, it then becomes pertinent to state that the rates of remediation will be reduced by sorption and/or diffusion into impermeable regions, with the overall rate controlled by the slowest process of desorption or biotransformation. For example, in batch systems where the solids have a large sorption capacity, only a small fraction of the contaminant mass may be present in the bulk water. Evolving experimental reports is thus emphasizing that sorbate *diffusion* is often the limiting step, particularly in systems involving contaminants with large organic-carbon partition coefficients (k_{OC}) and large fine-pored aggregates with high organic matter content.

Studies on mass transfer effects for the evaluation of bioavailability and biodegradation parameters of contaminant solutes in aqueous solid/sediment matrix applicable for biodegradation of the environmentally persistent and recalcitrant chemicals have been extensively investigated using models [7], [9] which incorporates a two-site sorption/desorption kinetics.

The applications of mass transfer by diffusion in multi-component mixtures were carried out by comparing the adsorption and desorption behavior of polycyclic aromatic hydrocarbons and benzene, toluene, ethylbenzene and xylene (BTEX) in sand and clay sediments fractions. Using equilibrium time as a basis for argument, research result showed that the contaminants in sand attained equilibrium faster than in the clay sediment for both sorption and desorption studies as a result of the higher permeability of sand sediment. The desorption equilibrium time in both sediment types was found to be slower than adsorption an indication that it may be the limiting step in the event of mineralization.

The adsorption and desorption kinetics of naphthalene using calcined and modified clay soil fractions at ambient temperature has also been investigated to demonstrate the role of mass transfer in the removal of toxic chemicals from soil. Result of the batch experiments showed that adsorption equilibrium was attained at 24, 28 and 32 hours for modified, calcined and untreated soils; while the desorption equilibrium occurred at 46 hours for modified and 52 hours for calcined and untreated soils. Following the equilibrium time, the percentage of an initial 100mg/l of naphthalene in the slurry phase system unadsorbed was 30%, 32% and 35% for the calcined, modified and untreated soils respectively; while 12.3%, 11.2% and 9.5% of the adsorbed naphthalene resisted desorption by the calcined, modified and untreated samples respectively at equilibrium. The mass transfer rate was estimated using the Lagergren equation and found to be 0.12 mg/g.hr, 0.11mg/g.hr and 0.08mg/g.hr during adsorption while the desorption rate (k_{des}) were 0.06mg/g.hr, 0.05mg/g.hr and 0.07mg/g.hr for calcined, modified and untreated soil respectively. The results clearly confirm that the rate of adsorption and desorption of naphthalene differ among soil types and this directly affects its effective removal from the soil. Studies on the dynamic behavior of the adsorption of naphthalene onto natural clay with various modifying agents have shown that adsorption increased to a large extent with increasing surface area and porosity of the modified clay. The results using acids, bases and salts showed that equilibrium adsorption of naphthalene from the bulk solution were attained at a faster rate using inorganic acids. Further investigations revealed that of all the acids used, the diffusivity of naphthalene was highest in the HCl-modified clay which had the largest surface area and porosity.

Comparing this result with the data from the desorption study showed that while adsorption was inversely proportional to the pH of the medium, desorption was directly proportional. The implication of the retention time obtained from the equilibrium study is significant as it provides the bench mark for the interplay between sorption and degradation for transport and transformation of contaminant solutes within the soil matrix.

The adsorption and desorption kinetics of naphthalene, anthracene, and pyrene in a soil slurry reactor at ambient conditions was also investigated with a view to ascertaining the mechanisms controlling the retention and release rates of the compounds in the soil matrix. A stirred-flow method was employed to perform the experiments [10].

Analysis of the results of the batch adsorption/desorption kinetics and equilibria indicated that the desorption rate was slower than the adsorption rate. The cumulative extent of desorption for the tested chemicals (naphthalene, anthracene and pyrene) suggested that the desorption step was also the rate limiting for biodegradation. This may not be unrelated to the fact that diffusion in the pores may have been retarded by surface adsorption effects on soil organic carbon. The observed resultant effect was the lowering of the aqueous phase concentration of the contaminant PAHs which renders them not readily available to the microorganisms. The biodegradation will thus in the long run be controlled by the slow desorptive and diffusive mass transfer into biologically active areas. The extent of partitioning for the polycyclic aromatic hydrocarbons tested was found to be dependent on their solubility and diffusivity in the aqueous phase see Table 3. Experimental results from this study affirm that diffusivity/diffusion coefficient is a property dependent on the physical properties of a system as well as the molecularity, structural configuration and angularity of the solute [8].

Properties	Naphthalene	Anthracene	Pyrene
Molecular formula	$C_{10}H_8$	$C_{14}H_{10}$	$C_{16}H_{10}$
Molecular weight (g/mol)	128	178	202
Density (g/cm³)	1.14	1.099	1.271
Melting point (°C)	80.5	217.5	145-148
Boiling point (°C)	218	340	404
Aqueous solubility (g/m³)	0.93	0.07	0.14

*Adapted from Zander et al., (1993), Perry's Handbook of Chemical Engineers' (1998) and Oleszczuk and Baran, (2003).

Table 3. Some properties of investigated PAHs*

Overall, discussion of the subject under consideration indicates that mass transfer is assessed by the rate of adsorption/desorption and it must be emphasized that this is both the measured concentration of the solute with time and the mechanism of distributing the solutes into surfaces and pores of individual adsorbent medium. The successful predictions of the fate and transport of solutes in the environment is hinged on the availability of accurate transport parameters.

Solute transport with linear equilibrium therefore must as a matter of necessity be accounted for and be accommodated as an integral component of the mineralization of toxic chemicals.

Author details

Engr Owabor*

Address all correspondence to: owabor4you@yahoo.com

Department of Chemical Engineering, University of Benin, Benin City, Nigeria

References

[1] Treybal, R. E. (1981). Mass-Transfer Operations. , 19-23.

[2] Owabor, C. N, & Audu, J. E. (2010). Studies on the Adsorption of Naphthalene and Pyrene from aqueous medium using ripe orange peels as Adsorbent" Global Journal of Pure and Applied Sciences. , 16(1)

[3] Owabor, C. N, & Owhiri, E. (2011). Utilization of periwinkle shell carbon in the removal of polycyclic aromatic hydrocarbons from refinery wastewater." Journal of the Nigerian Society of Chemical Engineers, , 26

[4] Owabor, C. N, Ono, U. M, & Isuekevbo, A. (2012). Enhanced sorption of Naphtha-lene onto a Modified Clay Adsorbent: Effect of Acid, Base and Salt modifications of clay on sorption kinetics." Advances in Chemical Engineering and Science, , 2(3)

[5] Ehlers, J L, & Luthy, G R. (2003). Contaminant bioavailability in soil and sediment. Environmental Science and Technology, 37: 295A-302A.

[6] Bengtsson, G, & Carlsson, C. (2001). Degradation of dissolved and sorbed 2,4-di-chlorophenol in soil columns by suspended and sorbed bacteria. Biodegradation, , 12(6), 419-32.

[7] Owabor, C. N, Ogbeide, S. E, & Susu, A. A. (2010). Degradation of polycyclic aromat-ic hydrocarbons: Model simulation for bioavailability and biodegradation". Canadi-an Journal of Chemical Engineering. , 88(2)

[8] Owabor, C. N. (2007). PhD Thesis, University of Lagos, Nigeria.

[9] Haws, N. W, Ball, P. W, & Bouwer, E. J. (2006). Modeling and interpreting bioavaila-bility of organic contaminant mixtures in subsurface environments". J. of Contami-nant Hydrology , 82, 255-292.

[10] Owabor, C. N, Ogbeide, S. E, & Susu, A. A. (2010). Adsorption and Desorption kinet-ics of Naphthalene, Anthracene and Pyrene in soil matrix". Petroleum Science and Technology. , 28(5)

[11] Oladele, A. O. Thesis. University of Benin, Nigeria.

[12] Zander et al(1993). Physical. and chemical properties of polycyclic aromatic hydro-carbons. Handbook of Polycyclic Aromatic Hydrocarbons. Marcel Dekker, Inc., New York, , 1-26.

[13] Perry, R. H, & Green, D. W. (1998). Perry's Chemical Engineers Handbook, 7th Edi-tion. McGraw-Hill Inc.

[14] Oleszczuk and Baran(2003). Degradation of individual polycyclic aromatic hydrocar-bons (PAHs) in soil polluted with aircraft fuel." Polish J. Environ. Studies, , 12, 431-437.

Chapter 10

Coordinated and Integrated Geomorphologic Analysis of Mass Transfers in Cold Climate Environments – The SEDIBUD (Sediment Budgets in Cold Environments) Programme

Achim A. Beylich

Additional information is available at the end of the chapter

1. Introduction

General background

Geomorphologic processes, responsible for transferring sediments and effecting landform change, are highly dependent on climate, and it is anticipated that climate change will have a major impact on the behaviour of Earth surface systems. Research on sedimentary fluxes from source to sink in a variety of different climatic environments is represented by a substantial body of literature. Studies on source-to-sink fluxes generally refer to the development of sediment budgets. A sediment budget is an accounting of the sources and disposition of sediment as it travels from its point of origin to its eventual exit from a defined landscape unit like a drainage basin, e.g. [1]. Accordingly, the development of a sediment budget necessitates the identification of processes of erosion, transport and deposition within a defined area, and their rates and controls [1, 2, 3]. The fundamental concept underpinning source-to-sink sediment flux and sediment budget studies is the basic sediment mass balance equation:

$I = O +\text{-}\Delta S$

Where inputs (I) equal outputs (O) plus changes in net storage of sediment (ΔS). Source-to-sink studies permit quantification of the transport and storage of sediment in a system. A thorough understanding of the current sediment production and flux regime within a system is fundamental to predict likely effects of changes to the system, whether climatic induced or human-influenced. Source-to-sink sediment flux and sediment budget research therefore enables the prediction of changes to erosion and sedimentation rates, knowledge of where sediment will be deposited, how long it will be stored and how much sediment will be remobilised [1, 3, 4].

Sediment sources

Sediments are eroded and mobilised in source areas. Sediment sources are diverse and subject to variation in response to climate change. Global warming leads to the loss of glacial ice, which in turn increases slope instability caused by glacial de-buttressing, and flooding from glacial and moraine-dammed lakes [5, 6]. All these processes redistribute sediments and operate at different rates as a result of change to the system. Glaciers and ice sheets exert strong controls on the supply of sediments. For example, Knight et al. [7] identify the basal ice layer of a section of the Greenland ice sheet as the dominant source of sediment production. There is, however, only limited knowledge of debris fluxes from ice sheets and glaciers and its variability. The main mechanisms of sediment production in source areas can be described in terms of contemporary environmental conditions. However, in order to fully understand sediment supply a longer-term perspective is needed. Over the Quaternary, glacier fluctuations have had profound influences in depositing extensive mantles of sediments. More-widely, periglacial activity has altered the landscape under non-glacial cold climate conditions. The obvious imprint of this legacy is often reflected in contemporary sediment transfer rates where pre-existing deposits are eroded by present-day processes [6, 8].

Sediment transfers

Sediment transfers move eroded sediments from their source area to an area of temporal storage or long-term deposition in sinks. Rates of sediment transfer are not only conditioned by competence of geomorphic processes but also by the availability of sediment for transport. Accordingly, in assessing sediment transfer we need to quantify the forces, which drive transport processes but equally account for the factors, which control sediment supply [8]. Glacial fluxes are arguably the most significant processes for contemporary sediment flux [9]. Small-scale process studies very often focus on sedimentary fluxes from areas of weathering and erosion to areas of storage within defined landscape units like drainage basins, whereas large-scale sediment systems couple headwaters to oceanic sinks. For example, Gordeev [10], applying models developed by Morehead et al. [11], estimates the increase in sediment load in Arctic rivers in response to a rise in surface temperature of the drainage basins. Based on this model, increases in river discharge lead to an increase in the sediment flux of the six largest Arctic rivers, predicted to range from 30% to 122% by the year 2100.

Sediment stores / sinks

The identification of storage elements and sinks is critical to the effective study and understanding of source-to-sink sedimentary fluxes [1]. The setting of a particular drainage basin defines the boundary conditions for storage within that landscape unit. Within a defined landscape unit like a drainage basin, the slope and valley infill elements constitute the key storage units and storage volumes are important for addressing time-dependent sediment budget dynamics. Dating of storage in sedimentary source-to-sink flux studies is applied to determine or estimate the ages and chronology of the storage components within the system. An understanding of the nature of primary stores, secondary stores and the potential storage capacities of different types of drainage basins is important along with knowledge of sediment residence times. Of growing importance is the development of innovative field methods, such as geophysical techniques for estimating sediment storage volumes [12, 13, 14]. Within large-scale sediment systems oceanic sinks are most important and provide the opportunity to estimate rates of sediment production and delivery at long-term temporal as well as continental spatial scales [15, 16].

2. The I.A.G. / A.I.G. SEDIBUD programme

Amplified climate change and ecological sensitivity of polar and cold environments has been highlighted as a key global environmental issue [17]. Projected climate change in cold regions is expected to alter melt season duration and intensity, along with the number of extreme rainfall events, total annual precipitation and the balance between snowfall and rainfall. Similarly, changes to the thermal balance are expected to reduce the extent of permafrost and seasonal ground frost and increase active layer and thaw depths. These effects will undoubtedly change surface environments in cold environments and alter the fluxes of sediments, nutrients and solutes, but the absence of data and analysis to understand the sensitivity of the surface environment are acute in cold climate environments.

The *SEDIBUD (Sediment Budgets in Cold Environments)* Programme of the International Association of Geomorphologists (I.A.G./A.I.G.) was formed in 2005 to address this identified key knowledge gap [18, 19]. SEDIBUD currently has about 400 members worldwide and the Steering Committee of this international programme is composed of ten scientists from eight different countries:

- Achim A. Beylich (*Chair*) (Norway)

- Armelle Decaulne (*Secretary*) (France)

- John C. Dixon (USA)

- Scott F. Lamoureux (*Vice-Chair*) (Canada)

- John F. Orwin (Canada)

- Jan-Christoph Otto (Austria)

- Irina Overeem (USA)

- Þorsteinn Sæmundsson (Iceland)

- Jeff Warburton (UK)

- Zbigniew Zwolinski (Poland)

The central research question of this global group of scientists is to

Assess and model the contemporary sedimentary fluxes in cold climates, with emphasis on both particulate and dissolved components.

Initially formed as European Science Foundation (ESF) Network SEDIFLUX (2004-2006) [20, 21], SEDIBUD has further expanded to a global group of researchers with in total 44 field research sites (SEDIBUD Key Test Sites) located in polar and alpine regions in the northern and southern hemisphere, see [22]. Research carried out at each site varies by programme, logistics and available resources, but typically represents interdisciplinary collaborations of geomorphologists, hydrologists, ecologists, permafrost scientists and glaciologists. SEDIBUD has developed a key set of primary surface process monitoring and research data requirements to incorporate results from these diverse projects and allow coordinated quantitative analysis across the programme. SEDIBUD Key Test Sites provide data on annual climate conditions, total discharge and particulate and dissolved fluxes as well as information on other relevant surface processes. A number of selected SEDIBUD Key Test Sites is providing high-resolution data on climate conditions, runoff and sedimentary fluxes, which in addition to the annual data contribute to the SEDIBUD Metadata Database which is currently developed. To support these coordinated efforts, the SEDIFLUX Manual [3] has been produced to establish common methods and data standards [18, 19]. In addition, a framework paper for characterizing fluvial sediment fluxes from source to sink in cold environments has been published by the group [23].

Comparable datasets from different SEDIBUD Key Test Sites are analysed to address key research questions of the SEDIBUD Programme as defined in the SEDIBUD Working Group Objective [24].

3. Compiled annual data from defined SEDIBUD key test sites

Table 1 compiles key parameters of four selected SEDIBUD research field sites (Figures 1 and 2) as examples.

SEDIBUD Key Test Site Catchment	Geographical coordinates	Area (km²)	Elevation range (m); Topographic relief (m)	Mean annual air temperature in °C	Annual precipitation (mm)	Lithology
Hrafndalur (Iceland)	65°28´N, 13°42´W	7	6 – 731; 725	3.6	1719	Rhyolites
Austdalur (Iceland)	65°16´N, 13°48´W	23	0 – 1028; 1028	3.6	1431	Basalt
Latnjavagge (Sweden)	68°20´N, 18°30E	9	950 – 1440; 490	-2.0	852	Mica-garnet schists
Kidisjoki (Finland)	69°47´N, 27°05´E	18	75 – 365; 290	-2.0	415	Gneisses and granulites

Table 1. Key parameters of four selected catchment geo-systems (SEDIBUD Key Test Sites) in Eastern Iceland, Swedish Lapland and Finnish Lapland.

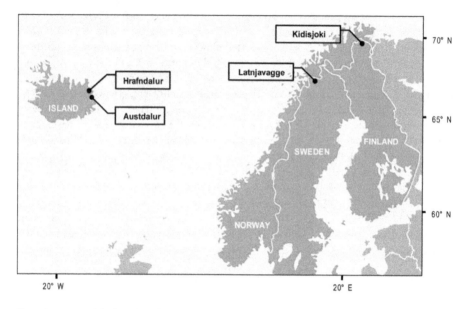

Figure 1. Location of the four selected SEDIBUD Key Test Sites Hrafndalur (Iceland), Austdalur (Iceland), Latnjavagge (Sweden) and Kidisjoki (Finland)

The generation and compilation of directly comparable data sets from the defined SEDIBUD Key Test Sites in the SEDIBUD Metadata Database is the basis for modelling effects of climate change on sedimentary fluxes and yields in cold climate environments by using space-for-time substitution [3, 18-21].

Annual data (as required from defined SEDIBUD Key Test Sites) from the four examples Hrafndalur (Iceland) [25, 26, 27], Austdalur (Iceland) [26, 27], Latnjavagge (Sweden) [25, 26, 28] and Kidisjoki (Finland) [25, 26] are compiled in Table 2. Time series of these mean annual data are published in [25-28].

4. Direct comparison and major controls of annual mass transfers within the four selected catchment geo-systems

On the basis of geomorphic process rates which were calculated for the Hrafndalur, Austdalur, Latnjavagge and Kidisjoki drainage basins after longer-term field studies (several years of process monitoring, mapping and observation) [26], the absolute and the relative importance of present-day denudative surface processes in the entire catchments was estimated by the quantification of the mass transfers caused by the different denudative surface processes. To allow direct comparison of the different denudative processes, all mass transfers are shown as tonnes multiplied by meter per year (t m yr^{-1}), i.e. as the product of the annually transferred mass and the corresponding transport distance, see [26, 29-31].

Coordinated and Integrated Geomorphologic Analysis of Mass Transfers in Cold Climate
Environments – The SEDIBUD (Sediment Budgets in Cold Environments) Programme

235

Hrafndalur

Austdalur

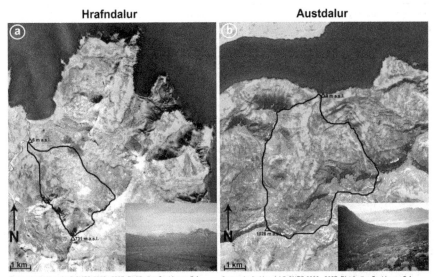

Source: Inniheldur efni © CNES 2002 - 2007, Distribution Spot Image S.A.,
France, öll réttindi áskilin / © Landmælingar Íslands 2011

Source: Inniheldur efni © CNES 2002 - 2007, Distribution Spot Image S.A.,
France, öll réttindi áskilin / © Landmælingar Íslands 2011

Latnjavagge

Kidisjoki

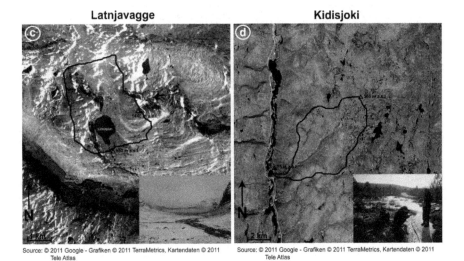

Source: © 2011 Google - Grafiken © 2011 TerraMetrics, Kartendaten © 2011
Tele Atlas

Source: © 2011 Google - Grafiken © 2011 TerraMetrics, Kartendaten © 2011
Tele Atlas

Figure 2. Views of the four selected SEDIBUD Key Test Sites Hrafndalur (Eastern Iceland), Austdalur (Eastern Iceland),
Latnjavagge (Swedish Lapland) and Kidisjoki (Finnish Lapland)

Name of SEDIBUD Key Test Site: Hrafndalur (Iceland) Principal Investigator: Achim A. Beylich	Period of investigations (years): 2002 - 2010 (Hydrological Year (HY) or Calender Year (CY); Published Data (PD) or Unpublished Data (UPD))
Mean annual temperature (°C):	3.6
Total annual precipitation [mm]:	1719
Total annual runoff [mm]:	1344
Annual suspended sediment yield [t km^{-2}]:	19
Annual solute yield (atmospherically corrected) [t km^{-2}]:	29
Name of SEDIBUD Key Test Site: Austdalur (Iceland) Principal Investigator: Achim A. Beylich	Period of investigations (years): 1996 - 2010 (Hydrological Year (HY) or Calender Year (CY); Published Data (PD) or Unpublished Data (UPD))
Mean annual temperature (°C):	3.6
Total annual precipitation [mm]:	1431
Total annual runoff [mm]:	1130
Annual suspended sediment yield [t km^{-2}]:	42
Annual solute yield (atmospherically corrected) [t km^{-2}]:	8
Name of SEDIBUD Key Test Site: Latnjavagge (Sweden) Principal Investigator: Achim A. Beylich	Period of investigations (years): 2000 - 2010 (Hydrological Year (HY) or Calender Year (CY); Published Data (PD) or Unpublished Data (UPD))
Mean annual temperature (°C):	-2.0
Total annual precipitation [mm]:	852
Total annual runoff [mm]:	717
Annual suspended sediment yield [t km^{-2}]:	2.4
Annual solute yield (atmospherically corrected) [t km^{-2}]:	4.9
Name of SEDIBUD Key Test Site: Kidisjoki (Finland) Principal Investigator: Achim A. Beylich	Period of investigations (years): 2002 - 2010 (Hydrological Year (HY) or Calender Year (CY); Published Data (PD) or Unpublished Data (UPD))
Mean annual temperature (°C):	-2.0
Total annual precipitation [mm]:	415
Total annual runoff [mm]:	324
Annual suspended sediment yield [t km^{-2}]:	0.3
Annual solute yield (atmospherically corrected) [t km^{-2}]:	3.1

Table 2. Compiled annual data from four selected SEDIBUD Key Test Sites

As based on these quantitative investigations, in all selected study areas in sub-Arctic oceanic eastern Iceland, Arctic oceanic Swedish Lapland and sub-Arctic oceanic Finnish Lapland the intensity of contemporary denudative surface processes and mass transfers caused by these geomorphic processes is altogether rather low.

A direct comparison of the annual mass transfers within the four investigated drainage basins (Figure 3) summarises that there are differences between process intensities and the relative importance of different denudative processes within the study areas in Eastern Iceland, Swedish Lapland and Finnish Lapland.

The major controls of the detected differences are (see Figure 3):

i.   *Hydro-climate and connected runoff:*

The higher annual precipitation along with the larger number of extreme rainfall events and the higher frequency of snowmelt and rainfall generated peak runoff events in eastern Iceland as compared to Swedish Lapland and Finnish Lapland leads to higher mass transfers (Figure 3). All four study areas are located in oceanic cold regions and projected climate change is expected to alter melt season duration and intensity, along with an increased number of extreme rainfall events, total annual precipitation and the balance between snowfall and rainfall. In addition, changes in the thermal balance are expected to reduce the extent of permafrost and seasonal ground frost and increase active layer depths [17]. Looking at the existing differences between Hrafndalur / Austdalur (eastern Iceland), Latnjavagge (Swedish Lapland) and Kidisjoki (Finnish Lapland) it seems obvious that the projected changes in climate will cause significant changes of mass transfers.

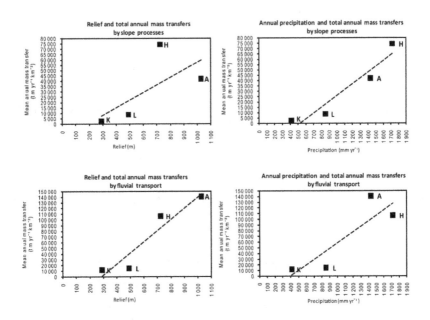

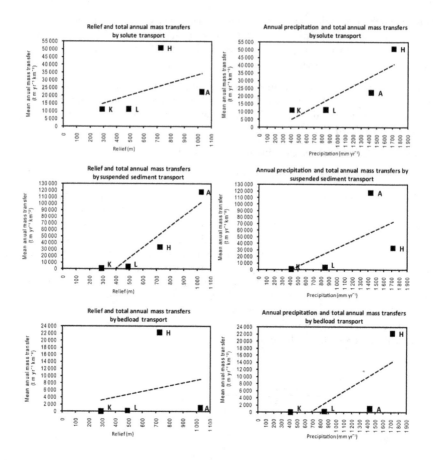

Figure 3. Statistical correlations between topographic relief and annual precipitation and annual mass transfers by slope processes and fluvial transport (fluvial solute transport, fluvial suspended sediment transport, fluvial bedload transport) for the four selected SEDIBUD Key Test Sites Hrafndalur (H), Austdalur (A), Latnjavagge (L) and Kidisjoki (K)

ii.   *Topographic relief:*

The greater steepness of the Icelandic drainage basins leads to larger mass transfers here as compared to Latnjavagge and especially to Kidisjoki (Figure 3).

iii.   *Lithology:*

The low resistance of the rhyolites in Hrafndalur causes especially high weathering rates and connected mass transfers in this drainage basin (see Figure 3). Due to the lower resistance of the rhyolites as compared to the basalts found in Austdalur Postglacial modification of the glacially formed relief is clearly further advanced in Hrafndalur as compared to Austdalur.

iv.   *Vegetation cover (with vegetation cover being partly modified by human activity):*

The significant disturbance of the vegetation cover by direct human impacts in Hrafndalur / Austdalur (eastern Iceland) causes higher mass transfers by slope wash here whereas restricted sediment availability is a major reason for lower mass transfers in Latnjavagge (Swedish Lapland) and Kidisjoki (Finnish Lapland).

5. Conclusions

As a result, hydro-climate and topographic relief, followed by lithology and vegetation cover (with vegetation cover being partly modified by human activity), are the main controls of the mass transfers modifying the investigated sub-Arctic / Arctic landscapes, see also [32]. More studies to the present one, carried out within the SEDIBUD Programme with unified geomorphologic field methods [3, 21, 33, 34] in environments having different climatic, vegetation, human impact, topographic, lithological / geological and/or tectonic features will help to gain improved understanding of the internal differentiation of different global cold climate environments, see e.g. [21, 33, 35, 36]. Furthermore, additional information on the control mechanisms of processes, the role of extreme geomorphic events for longer-term mass transfers and sediment budgets, the general intensity of geomorphic processes and mass transfers, and the relative importance of different processes for slope and valley formation and relief development under different environmental conditions can be collected. Direct comparisons of SEDIBUD Key Test Sites (catchment geo-systems) and the application of the Ergodic principle of space-for-time substitution will improve the possibilities to model relief development as well as possible effects of projected climate change in cold climate environments.

Acknowledgements

The author would like to thank the SEDIBUD Steering Committee Members and numerous SEDIBUD Members for fruitful discussions and numerous valuable inputs.

Author details

Achim A. Beylich

Address all correspondence to: achim.beylich@ngu.no

Geological Survey of Norway (NGU), Geo-Environment Division, Trondheim, Norway

References

[1] Reid L.M., Dunne T. Rapid evaluation of sediment budgets. Catena 1996.

[2] Slaymaker O. Research developments in the hydrological sciences in Canada (1995-1998): Surface water – quantity, quality and ecology. Hydrological Processes 2000; 14 1539-1550.

[3] Beylich A.A., Warburton, J. editors. Analysis of source-to-sink fluxes and sediment budgets in changing high-latitude and high-altitude cold environments. SEDIFLUX Manual. NGU Report 2007.053.

[4] Gurnell A.M., Clark M.J. editors. Glacio-fluvial sediment transfer: An Alpine perspective. Chichester: Wiley; 1987.

[5] Evans S.G., Clague J.J. Recent climate change and catastrophic geomorphic processes in mountain environments. Geomorphology 1994; 10 107-128.

[6] Ballantyne C.K. Paraglacial geomorphology. Quaternary Science Reviews 2002; 21 1935-2017.

[7] Knight P.G., Waller R.I., Patterson C.J., Jones A.P., Robinson Z.P. Discharge of debris from ice at the margin of the Greenland ice sheet. Journal of Glaciology 2002; 48 192-198.

[8] Warburton J. Sediment budgets and rates of sediment transfer across cold environments in Europe: a commentary. Geografiska Annaler 2007; 89A(1) 95-100.

[9] Harbor J., Warburton J. Glaciation and denudation rates. Nature 1992; 356 751.

[10] Gordeev V.V. Fluvial sediment flux to the Arctic Ocean. Geomorphology 2006; 80 94-104.

[11] Morehead M.D., Syvitski J.P., Hutton E.W., Peckham S.D. Modeling the temporal variability in the flux of sediment from ungauged river basins. Global and Planetary Change 2003; 39 95-110.

[12] Schrott L., Hufschmidt G., Hankammer M., Hoffmann T., Dikau R. Spatial distribution of sediment storage types and quantification of valley fill deposits in an alpine basin, Reintal, Bavarian Alps, Germany. Geomorphology 2003; 55 45-63.

[13] Sass O. Spatial patterns of rockfall intensity in the northern Alps. Zeitschrift für Geomorphologie N.F. 2005; Suppl. 138 51-65.

[14] Hansen L., Beylich A.A., Burki V., Eilertsen R., Fredin O., Larsen E., Lyså A., Nesje A., Stalsberg K., Tønnesen J.-F. Stratigraphic architecture and infill history of a deglaciated bedrock valley based on georadar, seismic profiling and drilling. Sedimentology 2009; 56 1751-1773.

[15] Rise L., Ottesen D., Berg K., Lundin E. Large-scale development of the mid-Norwe-
      gian margin during the last 3 million years. Marine and Petroleum Geology 2005; 22
      33-44.

[16] Dowdeswell J.A., Ottesen D., Rise L. Flow switching and large-scale deposition by
      ice streams draining former ice sheets. Geology 2006; 34 313-316.

[17] ACIA, Impacts of a warming Arctic: Arctic Climate Impact Assessment. ACIA Over-
      view Report. Cambridge University Press; 2004.

[18] Beylich A.A. Quantitative studies on sediment fluxes and sediment budgets in
      changing cold environments – potential and expected benefit of coordinated data ex-
      change and the unification of methods. Landform Analysis 2007; 5 9-10.

[19] Beylich A.A., Lamoureux S.F., Decaulne A. Coordinated quantitative studies on sedi-
      ment fluxes and sediment budgets in changing cold environments examples from
      three SEDIBUD key test sites in Canada, Iceland and Norway. Landform Analysis
      2007; 5 11-12.

[20] Beylich A.A., Etienne S., Etzelmüller B., Gordeev V.V., Käyhkö J., Rachold V., Russell
      A.J., Schmidt K.-H., Sæmundsson Th., Tweed F.S., Warburton J. Sedimentary Source-
      to-Sink-Fluxes in Cold Environments – Information on the European Science Foun-
      dation (ESF) Network SEDIFLUX. Zeitschrift für Geomorphologie N.F. 2005; Suppl.
      138 229-234.

[21] Beylich A.A., Etienne S., Etzelmüller B., Gordeev V.V., Käyhkö J., Rachold V., Russell
      A.J., Schmidt K.-H., Sæmundsson Th., Tweed F.S., Warburton J. The European Sci-
      ence Foundation (ESF) Network SEDIFLUX – An introduction and overview. Geo-
      morphology 2006; 80(1-2) 3-7.

[22] Beylich A.A., Decaulne A., Dixon J.C., Lamoureux S.F., Orwin J.F., Otto J.-Ch., Over-
      eem I., Sæmundsson Th., Warburton J., Zwolinski Z. The global Sediment Budgets in
      Cold Environments (SEDIBUD) Programme: Coordinated studies of sedimentary
      fluxes and budgets in changing cold environments. Zeitschrift für Geomorphologie
      2012; 56(1) 3-8.

[23] Orwin J.F., Lamoureux S.F., Warburton J., Beylich A.A. A framework for characteriz-
      ing fluvial sediment fluxes from source to sink in cold environments. Geografiska
      Annaler 2010; 92A(2) 155-176.

[24] International association of Geomorphologists (I.A.G. / A.I.G.): SEDIBUD Website.
      http://www.geomorph.org/wg/wgsb.html (accessed 9 June 2012).

[25] Beylich A.A. Chemical and mechanical fluvial denudation in cold environments –
      Comparison of denudation rates from three catchments in sub-Arctic Eastern Ice-
      land, sub-Arctic Finnish Lapland and Arctic Swedish Lapland. Jökull 2009; 59 19-32.

[26] Beylich A.A. Mass transfers, sediment budgets and relief development in cold envi-
      ronments: Results of long-term geomorphologic drainage basin studies in Iceland,

Swedish Lapland and Finnish Lapland. Zeitschrift für Geomorphologie 2011; 55(2) 145-174.

[27]   Beylich A.A., Kneisel Ch. Sediment budget and relief development in Hrafndalur, sub-Arctic oceanic Eastern Iceland. Arctic, Antarctic and Alpine Research 2009; 41(1) 3-17.

[28]   Beylich A.A., Mass transfers, sediment budget and relief development in the Latnja-vagge catchment, Arctic oceanic Swedish Lapland. Zeitschrift für Geomorphologie N.F. 2008; 52(1) 149-197.

[29]   Jäckli H. Gegenwartsgeologie des Bündnerischen Rheingebietes. Beitrag zur Geologischen Karte der Schweiz. Geotechnische Serie 1957; 36.

[30]   Rapp A. Recent development of mountain slopes in Kärkevagge and surroundings, Northern Scandinavia. Geografiska Annaler 1960; 42 71-200.

[31]   Barsch D. Studien zur gegenwärtigen Geomorphodynamik im Bereich der Oobloyah Bay, N-Ellesmere Island, N.W.T., Kanada. Heidelberger Geographische Arbeiten 1981; 69 123-161.

[32]   Slaymaker O., Spencer T., Dadson S. Landscape and landscape-scale processes as the unfilled niche in the global environmental change debate: an introduction. In: Slaymaker O., Spencer T., Embleton-Hamann C. (eds.) Geomorphology and Global Environmental Change. Cambridge University Press; 2009. p1-36.

[33]   Beylich A.A., Lamoureux S.F., Decaulne A. The global I.A.G./A.I.G. SEDIBUD (Sediment Budgets in Cold Environments) programme: Introduction and overview. Norwegian Journal of Geography 2008; 62(2) 50-51.

[34]   Beylich A.A., Lamoureux S.F., Decaulne A. Developing frameworks for studies on sedimentary fluxes and budgets in changing cold environments. Quaestiones Geographicae 2011; 30(1) 5-18.

[35]   Barsch D. Geomorphologische Untersuchungen zum periglazialen Milieu polarer Geosysteme. Zeitschrift für Geomorphologie N.F. 1984; Suppl. 50 107-116.

[36]   Barsch D. Forschungen in Polargebieten. Heidelberger Geowissenschaftliche Abhandlungen 1986; 6 33-50.

Permissions

The contributors of this book come from diverse backgrounds, making this book a truly international effort. This book will bring forth new frontiers with its revolutionizing research information and detailed analysis of the nascent developments around the world.

We would like to thank Hironori Nakajima, for lending his expertise to make the book truly unique. He has played a crucial role in the development of this book. Without his invaluable contribution this book wouldn't have been possible. He has made vital efforts to compile up to date information on the varied aspects of this subject to make this book a valuable addition to the collection of many professionals and students.

This book was conceptualized with the vision of imparting up-to-date information and advanced data in this field. To ensure the same, a matchless editorial board was set up. Every individual on the board went through rigorous rounds of assessment to prove their worth. After which they invested a large part of their time researching and compiling the most relevant data for our readers. Conferences and sessions were held from time to time between the editorial board and the contributing authors to present the data in the most comprehensible form. The editorial team has worked tirelessly to provide valuable and valid information to help people across the globe.

Every chapter published in this book has been scrutinized by our experts. Their significance has been extensively debated. The topics covered herein carry significant findings which will fuel the growth of the discipline. They may even be implemented as practical applications or may be referred to as a beginning point for another development. Chapters in this book were first published by InTech; hereby published with permission under the Creative Commons Attribution License or equivalent.

The editorial board has been involved in producing this book since its inception. They have spent rigorous hours researching and exploring the diverse topics which have resulted in the successful publishing of this book. They have passed on their knowledge of decades through this book. To expedite this challenging task, the publisher supported the team at every step. A small team of assistant editors was also appointed to further simplify the editing procedure and attain best results for the readers.

Our editorial team has been hand-picked from every corner of the world. Their multi-ethnicity adds dynamic inputs to the discussions which result in innovative outcomes. These outcomes are then further discussed with the researchers and contributors who give their valuable feedback and opinion regarding the same. The feedback is then collaborated with the researches and they are edited in a comprehensive manner to aid the understanding of the subject.

Apart from the editorial board, the designing team has also invested a significant amount of their time in understanding the subject and creating the most relevant covers. They scrutinized every image to scout for the most suitable representation of the subject and create an appropriate cover for the book.

The publishing team has been involved in this book since its early stages. They were actively engaged in every process, be it collecting the data, connecting with the contributors or procuring relevant information. The team has been an ardent support to the editorial, designing and production team. Their endless efforts to recruit the best for this project, has resulted in the accomplishment of this book. They are a veteran in the field of academics and their pool of knowledge is as vast as their experience in printing. Their expertise and guidance has proved useful at every step. Their uncompromising quality standards have made this book an exceptional effort. Their encouragement from time to time has been an inspiration for everyone.

The publisher and the editorial board hope that this book will prove to be a valuable piece of knowledge for researchers, students, practitioners and scholars across the globe.

List of Contributors

Rafał Rakoczy, Marian Kordas and Stanisław Masiuk
Institute of Chemical Engineering and Environmental Protection Process, West Pomeranian University of Technology, Poland

Fabien Beaumont and Guillaume Polidori
GRESPI/Thermomécanique, Université de Reims, France

Gérard Liger-Belair
GSMA, UMR CNRS 7331, Université de Reims, France

T.A. Vartanyan, N.B. Leonov, S.G. Przhibel'skii and N.A. Toropov
St. Petersburg National Research University of Information Technology, Mechanics and Optics, St. Petersburg, Russian Federation

Hironori Nakajima
Department of Mechanical Engineering at Kyushu University, Japan

Ana María Mendoza Martínez
Technological Institute of Madero City, Division of Graduate Studies and Research (ITCM), Madero, México

Eleazar Máximo Escamilla Silva
Chemical Departments, Technological Institute of Celaya, Celaya, México

Krzysztof Górnicki, Agnieszka Kaleta, Radosław Winiczenko, Aneta Chojnacka and Monika Janaszek
Faculty of Production Engineering, Warsaw University of Life Sciences, Poland

M. A. S. D. Barros and P. A. Arroyo
Department of Chemical Engineering, State University of Maringá, Maringá, Brazil

E. A. Silva
Department of Chemical Engineering, West Paraná State University, Jardim La Salle, Toledo, Brazil

A. Haghighi Asl and M. Khajenoori
School of Chemical Gas and Petroleum Engineering, Semnan University, Semnan, I.R., Iran

Engr Owabor
Department of Chemical Engineering, University of Benin, Benin City, Nigeria

Achim A. Beylich
Geological Survey of Norway (NGU), Geo-Environment Division, Trondheim, Norway